# BITPOCALYPSE

## RISE OF DIGITAL REBELLION

TOBIUS NELSON

*Bitpocalypse: Rise of the Digital Rebellion*

For licensing, media inquiries, or permissions, visit:
www.thbitpocalypse.com

ISBN: 979-8-9988791-0-4

First Printing, June 12, 2025

Printed in the United States of America

# DEDICATION

*To my family—Ashley, Spencer, Lincoln, and the one on the way—*
*You are my foundation. Your love supports me, your belief in me pushes me,*
*and your presence inspires me to become better every day.*
*This book exists because of you.*

# PREFACE

*Bitpocalypse* is more than a story—it's a reflection.

While the world of this novel is fictional, the ideas that power it are deeply personal. In many ways, this book is a mirror of my life: a fusion of my time in the fast-moving world of cryptocurrency, my career in technology, and my ever-growing desire to question the systems we live within.

As I built in crypto communities and worked behind the scenes in tech, I began to see patterns—fragments of a larger digital truth that felt too important to ignore. I wanted to channel those insights into something imaginative, something honest, something that could spark a new way of thinking.

That's how *Bitpocalypse* was born.

It's a story filtered through the lens of high technology, but grounded in very human truths: control, freedom, legacy, and resistance. It's not just science fiction—it's what I call *tech fiction*—a genre that blends emerging realities with the creative freedom of speculative storytelling.

I wrote this book for thinkers, for builders, for the rebels who see beyond the surface. I wrote it for my kids and their future. And I wrote it for anyone who believes stories still have the power to wake us up.

Welcome to the *Bitpocalypse*.

—Tobius Nelson
Atlanta, GA

# EPIGRAPH

"To build the future, you have to outpace your past
or you'll inherit its limits." — *Tobius Nelson*

# TABLE OF CONTENTS

# CHAPTER 1:
# DESTINED TO FAIL

"I've been running my whole life, but nothing could have prepared me for this final stand against GovCore."

I stand here with the weight of the world pressing on my shoulders and the cold steel of inevitability in my hands. The echoes of a lifetime of battles and turmoil ring in my ears. GovCore's relentless pursuit has finally brought me to this precipice, where the past and future collide in a blaze of defiance.

But before this moment, there was another time, a beginning shaped by struggle and uncertainty. Let me take you back to where it all started.

The hospital lights flickered, casting eerie shadows on the walls. My mother, Angel Valorian, was rushed through the sterile corridors on a gurney. Her face was contorted in pain and fear. The nurses moved with practiced urgency; their hushed voices were a stark contrast to the chaos within her body. She was having a stroke, and I was fighting to be born with an umbilical cord wrapped tightly around my neck.

"Doctor, her blood pressure is through the roof. We need to act fast!" a nurse shouted. Her hands trembled as she prepared the IV line.

Dr. Patel, a seasoned obstetrician, glanced at the monitors. "We're losing her. Prepare for an emergency C-section."

In the corner of the room, a man in a crisp suit observed with cold detachment. He was a GovCore agent, the hospital czar who oversaw the community with an iron fist. As Angel was wheeled into the operating room, he spoke into a small earpiece.

"Let it play out," he said. "The odds are one of them won't make it. Either way, it's a win for us."

Angel's grip on consciousness was slipping, memories of her life flashing before her eyes. She had moved from Monroe, GA, to Atlanta, hoping for a better future. Instead, she found herself alone, divorced, and facing the possible death of her unborn child.

As the surgical team worked frantically, Angel's mind drifted to the past. She remembered the days when she smoked cigarettes, given to her by well-meaning but misguided friends, unaware of the damage they could cause. She clung to the hope that her baby, despite the odds, would survive.

The room was tense with anticipation. Scalpel in hand, Dr. Patel made the incision. "We're almost there. Stay with us, Angel."

Minutes felt like hours as they fought to bring me into the world. Finally, a faint cry pierced the silence. I was born, against all odds, gasping for breath but alive.

My earliest memories are fragmented, hazy images of a life filled with struggle. Growing up in the rough neighborhoods of Atlanta, I quickly learned that survival was a daily battle. Our apartment on the south side was a haven for addicts and criminals, a place where police sirens were a nightly lullaby.

"Mom, why do we have to live here?" I asked one day. My voice was tinged with frustration.

Angel, still recovering from the stroke that had left her arm partially paralyzed, looked at me with sad eyes. "Toby, we don't have much choice. We have to make the best of it."

Poverty was a relentless companion that shadowed our every move. School was no refuge; I felt out of place; my gifted mind was stifled by rigid structures and low expectations. I often found myself hiding under the school steps, trying to escape the oppressive system that seemed determined to hold me back.

"He's brilliant, but he won't stay in class," my teacher would say. "We need to monitor him more closely."

It was during one of those days, hiding under the steps, that I first noticed them—men and women in dark suits, moving through the school with an air of authority that made the teachers visibly nervous. These were GovCore agents, and their presence was a clear indication that something was amiss.

One day, the school decided to test me to see if I was truly gifted. I was taken to a small room, away from the regular classes, and given a series of psychological evaluations and tests. The questions were challenging, but I breezed through them, feeling a sense of exhilaration as I solved every problem.

"You're doing great, Toby," my teacher, Mrs. Hargrove, said. Her voice was filled with pride. "Just a few more questions, and we'll be done."

After completing the test, she reviewed my answers. Her eyes widened with each passing moment. "You've passed with flying colors, Toby. You're one of the brightest students I've ever seen. You're a prodigy, a genius!"

Before I could fully absorb her words, the door burst open, and a GovCore agent stepped in. His cold, calculating eyes scanned the room and landed on Mrs. Hargrove. Without a word, he grabbed her arm and pulled her out of the room. I could hear muffled voices; the tension was palpable even through the closed door.

Minutes later, Mrs. Hargrove returned; her face was pale, and her hands were trembling. "I'm sorry, Toby," she said, her voice barely a whisper. "There's been a mistake. You didn't pass the test. You'll have to continue in your regular classes."

I was stunned. The truth was right in front of me: GovCore was overseeing every part of my life, ensuring that I remain in my place. The system was designed to suppress anyone who showed potential—to keep us from rising above our predetermined roles.

From that moment on, I felt like my every move was being watched, every success meticulously thwarted. But even then, a spark of defiance burned within me. I wouldn't let them break me, no matter how hard they tried.

As I grew older, the weight of our circumstances pressed heavier on my conscious. My mother's health continued to decline, and her struggles were compounded by the demands of raising me alone. Yet, through it all, she remained a pillar of strength; her faith was unshaken.

"I know it's hard, Toby, but you have to believe that things will get better," she would say. Her voice was filled with spiritual conviction.

Little did I know that these early trials were merely the prelude to the epic battle that lay ahead—a battle against a system determined to crush the human spirit. But for now, my story continues with the quiet resolve of a mother and the unyielding spirit of a boy destined to change the world.

After that fateful day at school, I was determined to uncover the truth about GovCore. I needed to understand who they were and why they had such a tight grip on my life. I scoured newspapers, books, and any scrap of information I could find about them. The more I learned, the more ominous their presence became.

GovCore was no ordinary organization. In the years following the major economic collapse, they had stepped in under the guise of restoring order. The government, weakened and desperate, had ceded control to this monolithic entity, believing it to be the only way to stabilize the country. But GovCore was not a savior; it was a corporate beast that thrived on control and manipulation.

The transformation of the government into a business-like entity was swift and brutal. Democracy gave way to a hierarchy reminiscent of a corporate pyramid, with GovCore at its apex. At the top were a select few—wealthy elites with bloodlines tracing back to old money and power. These individuals, untouchable and unaccountable, wielded absolute control over the corporation and, by extension, the nation.

As I delved deeper into the murky depths of GovCore's origins and operations, one chilling fact became glaringly clear: The leadership of GovCore was a tightly-knit web of interwoven bloodlines. The managers and directors, those who enjoyed the privileges just below the elite apex, were predominantly either born into or married into the select families that ruled the corporation.

"GovCore isn't just a corporation; it's a dynasty," I whispered to myself; the weight of the discovery pressed down on me.

The names on the organization chart that I had painstakingly pieced together began to blur into a tapestry of familial ties. The Johnsons, the Carters, the Kestrels—each name was a thread in the intricate fabric of control. It was as if the entire upper echelon of GovCore could be traced back to a handful of progenitors who had laid the foundation for this corporate behemoth.

"It's no coincidence they share the same last names," I noted. My voice was mingled with frustration and realization. "These aren't just random appointments; they're carefully orchestrated inheritances."

Marriages within these families were more than unions of love; they were strategic alliances that fortified the power structure of GovCore. The occasional outsider who ascended to these ranks inevitably did so through marriage, binding them to the core families. Wealth and power were passed down like heirlooms in this modern-day aristocracy.

"Climbing the ladder in GovCore, is not about what you know. It is all about your connections by blood or marriage," I concluded. There was a bitter taste in my mouth.

The realization that I was up against a multi-generational fortress of power and privilege was daunting. But it also fueled my resolve. This was the enemy, a cabal of names and faces bound by history and ambition, determined to maintain their grip on society.

"Bloodlines and last names—they're the chains that bind us to their rule. But chains can be broken," I thought, a spark of defiance ignited within me.

I discovered that while the pyramid-like structure of managers and directors enjoyed privileges, they were still beholden to the true power at the top. The vast majority, like myself, however, found themselves in the lower tiers, mere cogs in the sprawling machine, toiling endlessly to maintain the system that oppressed them.

The corporate structure was meticulously designed to perpetuate this inequality. The privileged few reaped the benefits of luxury and security while the rest of us labored to keep the system running. GovCore's reach extended into every facet of life, from the economy to education, and even into our homes.

As I continued my research, I noticed a disturbing pattern. GovCore's influence was pervasive and insidious. They controlled the media and shaped the narrative to suit their needs. Textbooks and encyclopedias were sanitized versions of history, glorifying GovCore's rise to power while erasing dissenting voices.

In the neighborhood, the presence of GovCore was a constant, oppressive reminder of our place in the hierarchy. Their sleek, imposing vehicles patrolled the streets with menacing regularity. Monthly—sometimes weekly—they rolled through. Dark windows hid the eyes that watched our every move.

I began to notice that their visits were often followed by incidents—crimes, disturbances, events that seemed almost orchestrated. And always, standing in the background, was a GovCore agent, observing, ensuring that the narrative of control remained unchallenged.

Their Inspections were terribly intrusive. Annually, GovCore officials would knock on every door, clipboard in hand, while conducting routine inspections of our homes. These visits were ostensibly safety checks, but everyone knew better. They were a reminder of who held the power, a subtle yet forceful way that kept us in line.

My mother, weary from years of hardship, would straighten the house, and her hands shook slightly as she opened the door to let them in. "Good morning, ma'am. Just a routine inspection," they would say, but their smiles never reaching their eyes. They would move through our small home, checking everything, noting every detail. It was a violation, a reminder that even within our walls, we were never truly safe from their gaze.

As I pieced together the information, a chilling realization settled over me. GovCore wasn't just a corporation; it was a surveillance state, a technological behemoth that monitored and controlled every aspect of our lives. Their power was absolute; their reach unending; and their intent was clear: to maintain the status quo, to ensure that only those deemed worthy could succeed.

But even in the face of such overwhelming power, a spark of defiance burned within me. I refused to be another cog in their machine. The knowledge I gained became my weapon, my shield against their insidious control. I was determined to fight, to resist in any way I could. I wanted to find a way to rise above the system that sought to crush me.

This was my reality, the world I was born into—a world where freedom was an illusion, and control was absolute. But I knew one thing for certain: I would not let GovCore dictate my destiny. My life was just beginning, and I was ready to challenge the system that sought to bind me.

One crisp autumn afternoon, a knock at the door announced the annual GovCore inspection. My mother's hands trembled slightly as she straightened her dress and opened the door. Standing there were two agents, a young woman and a man, both wearing the crisp, formal attire of GovCore officials. The woman, who introduced herself simply as Ela, had a disarming smile that didn't match her intentions.

"Good afternoon, ma'am," Ela said. Her voice was smooth and courteous. "We're here for your annual inspection."

"Of course," my mother replied, stepping aside to let them in. "Please, come in."

I never caught the name of the male agent who followed Ela into our modest home. His eyes swept over every detail with unsettling intensity. As they moved through the house, they separated, taking different sections while I watched them filled with curiosity and unease. There was something off about their presence, and that something set my nerves on edge.

Ela and the other agent moved through our home with an air of practiced efficiency, making notes and asking seemingly benign questions. But beneath their polite demeanor, I could sense a dark undercurrent. They weren't just inspecting. They were probing—searching for something specific.

As they reached the kitchen, Ela paused, and her gaze lingered on the table where my mother had placed a fresh load of laundry. She turned to my mother with an wavering smile. "I see you keep a tidy home, Mrs. Valorian. It's always a pleasure to see such care taken."

"Thank you," my mother replied. Her voice was tight with anxiety.

As the agents finished their inspection and prepared to leave, Ela's eyes met mine for a brief moment. There was a flicker of something—recognition, perhaps, or calculation. I couldn't be sure. They left without incident; their goodbyes were as polite and insincere as their greetings had been.

Later that evening, as I was playing with my cousins in the living room, we stumbled upon something that shouldn't have been there. A gun had been casually placed on the kitchen table as if it belonged there. My older cousin, intrigued and oblivious to the danger, picked it up.

"Hey, look at this! Is it real?" he asked, turning the weapon over in his hands.

"I don't know," I replied. My heart pounded in my chest. "Be careful with it."

We moved into another room, still curious and unaware of the true peril. My older cousin, thinking it was a toy, aimed the gun at me. "Bang, bang, you're dead!" he said with a grin.

Suddenly, a deafening boom echoed through the house. I felt a searing pain on the side of my head. My ears rang, and for a moment, I thought I was dead. Blood trickled down my temple, and I felt a burning sensation where the bullet had grazed me.

My mother's screams filled the air as she rushed into the room. "What happened? Oh my God, what happened?"

My cousin, pale and shaking, dropped the gun. "I didn't know it was real! I didn't know!"

Within minutes, the house was swarming with police and paramedics, and my cousin was taken away in handcuffs. His eyes were wide with fear and regret. As I was treated for my injury, I realized the full scope of what had happened.

This wasn't an accident. This was a setup. Ela and her partner hadn't come for an inspection—they had come to set a trap. They had left the gun, hoping it would lead to either my death or my incarceration. It was a deliberate move by a division of GovCore whose existence I would later uncover—a division dedicated to orchestrating scenarios designed to control and manipulate our destinies. They called

t "Destinatus," a name that would come to haunt me.

That day marked a turning point. My cousin's life was changed forever, and mine was, too. The innocence of childhood was ripped away and replaced by the harsh reality of GovCore's ruthless control. I knew then that they were watching me, monitoring my every move, and to be sure I stayed in my place, they would stop at nothing.

The seeds of rebellion that were planted that day were watered by the blood on my temple and the tears of my mother. I vowed to myself that I would not be their pawn. I would fight, resist, and one day, I would bring down the very system that sought to destroy me.

From the moment Ela and the other agent left our home that day, something shifted inside me. It was as though a veil had been lifted, revealing the dark, intricate web of control that GovCore had spun around my life. Every creak of the floorboards, every shadow that flickered past the window, sent a shiver down my spine. I became suspicious of everything and everyone. It was impossible to shake the feeling that I was being watched, that every step I took was being monitored and recorded by unseen eyes.

My curiosity turned into an obsession. I started paying close attention to the neighborhood, to the cars that drove by our house, to the faces that seemed to linger a little too long. The sleek, black vehicles with tinted windows became a regular sight on our street; their engines purred softly as they cruised past. I was always watching, always waiting.

"Did you see that car again, Mom?" I asked one day. My voice a mix of both fear and determination.

My mother glanced out the window, but she was too tired, too worn down by life to share my concern. "It's probably nothing, Toby. Just someone passing through."

But I knew better. I knew that GovCore was still out there, lurking in the shadows, waiting for the perfect moment to strike again.

I started keeping a notebook where I jotted down every detail—the make and model of the cars, the license plates, the number of times they passed by. I would hide in the bushes at the end of our block, watching as these mysterious vehicles circled our neighborhood like vultures waiting for something to die.

One afternoon, as I followed one of these cars on my bike, I noticed it pull into an alley behind a row of abandoned buildings. The driver got out, glanced around suspiciously, and then disappeared into one of the buildings. My heart raced as I approached, keeping a safe distance. I couldn't see what the man was doing inside, but I could hear the faint hum of machines, the clicking of keyboards, and the low murmur of voices.

"GovCore," I whispered to myself. The name automatically produced fear in my young mind.

I didn't dare go any closer, but I knew I had uncovered something important. Whatever they were doing in that building, it wasn't good. And it had everything to do with me.

As I became more entrenched in my surveillance of GovCore, a new figure entered our lives—Nate. My mother met him by chance one afternoon while struggling to carry groceries home from the store. Nate, who was tall and muscular, and displayed the disciplined bearing of a military man. He offered to help. His easy charm and confident demeanor quickly won her over.

"He's so kind, Toby," she told me with a smile one evening. "He reminds me of your father, but stronger, more protective."

At first, Nate seemed like a blessing. He helped around the house, fixed things that had been broken for years, and even made my mother laugh—a sound I hadn't heard in a long time. But something about him put me on edge. Maybe it was the way he looked at me, or the way he always seemed to know more about me than I had told him.

Nate started spending more time at our house, and before long, he and my mother were a couple. He moved in, and that's when things changed. The first sign had been the cigarettes. My mother had quit smoking years earlier, but Nate casually reintroduced her to the habit. "Just one or two won't hurt, Angel," he said, handing her a pack. "You deserve to relax."

Soon, the house was filled with the familiar acrid scent of smoke, a symbol of the comfort she thought she had found in Nate. But the smell mingled with something far more sinister as Nate's true nature began to surface.

The man who had Initially seemed like a protector quickly became a tyrant. He was quick to anger, and the smallest things would set him off. The first time he hit my mother, it was because dinner was a few minutes late. I'll never forget the sound of the sudden slap. It was like a gunshot in the night.

"Don't you ever make me wait again," he hissed. His voice was low and threatening.

My mother was shocked and hurt, but she simply nodded and apologized. I watched from the doorway with clinched fists. Rage bubbled up inside me. But I was just a kid, too small to do anything about it.

Despite the violence, there was another side to Nate—a side that intrigued me, even as it terrified me. He knew things about GovCore that no ordinary person should know. Late at night, after my mother had gone to bed, he would sit with me and talk about the world, about the forces that shaped it.

"You know, Toby . . ." he said one evening when the glow of his cigarette cast eerie shadows on the walls. "GovCore runs everything. They've got their hands in every pie, and they make sure people like you and me stay right where they want us."

"People like us?" I asked. I was confused.

"People who think too much, who see too much," he replied, exhaling a cloud of smoke. "They don't like that. They'll do whatever it takes to keep us in line."

As the months went by, Nate's true nature became increasingly apparent. He wasn't just a man with a temper—he was a strategist, a master manipulator who seemed to thrive on controlling every aspect of our lives, especially mine. At first, it was small things. If I left my shoes by the door instead of putting them away, or if I didn't clean my room to his exacting standards, he would explode.

"You think you can do whatever you want around here, Toby?" he would shout. His face was twisted with anger when he said, "Not in my house!"

And that's how he saw it—'his' house, 'his' rules. My mother, who had once been my shield, now stood by silently, too afraid or too entranced by Nate's influence to intervene. He twisted reality, making his harsh discipline seem like necessary tough love. He provided for us in ways we had never experienced before—a car, new clothes, even little luxuries like a big screen television, things we could not afford before he came along.

But those gifts came with a price. Nate used them to cement his authority, to create a dependency that made it almost impossible for my mother to see the truth. She was trapped in his web, and I was caught right there with her.

Whenever an opportunity arose for me to do something special—whether it was playing sports, participating in a school event, or even spending time with friends—Nate would find a way to snatch it away. The punishments were always severe, and they always came at the worst possible moments.

"You left your jacket on the floor again, Toby," Nate said one afternoon as he stood in the doorway of my room. His voice was deceptively calm. "You know what that means."

I knew all too well. It meant that the baseball game I'd looked forward to, the one I had trained for weeks to play in, was no longer happening for me. It meant that I would be grounded, forced to stay home and do chores under Nate's watchful eye.

"But it was hung up," I protested. Desperation crept into my voice. "I didn't leave it there."

Nate's eyes hardened. "It doesn't matter. Rules are rules. You need to learn discipline, Toby."

The way he said "discipline" made my skin crawl. It wasn't just about teaching me responsibility—it was about control, about keeping me from achieving anything that might make me stand out, anything that might give me a sense of accomplishment or self-worth.

One day, while cleaning the house as part of yet another punishment, I found something that confirmed my worst fears. Hidden in the back of my mother's closet was a small, leather-bound ledger. Inside were pages filled with notes— about me. Each entry meticulously documented my behavior, my interests, the things I excelled at, and the punishments Nate had doled out to keep me in line.

"October 3: Toby shows interest in joining the school chess club. Must ensure he's kept too busy with chores to attend."

"November 15: Toby excels in math. Recommended further reduction in access to study materials to avoid drawing attention at school."

"December 21: Missed opportunity to prevent participation in spelling bee. Need to tighten oversight." He wrote that about the school spelling bee, which I won.

My hands shook as I read through the ledger. It was clear that Nate was systematically dismantling any chance I had to succeed, to be anything more than what GovCore had planned for me. He wasn't just punishing me for minor infractions—he was ensuring that I never rose above mediocrity, that I stayed firmly under the radar, unnoticed and unremarkable.

But amid the turmoil, something unexpected happened. My mother became pregnant again. Despite the violence, despite the fear, she still found comfort in Nate, and that comfort led to the birth of my brother, Arthur. We would always call him Ace.

Ace was born into a world already shaped by GovCore's influence. From the moment he took his first breath, the forces that sought to control our lives were already at work. But he was my brother, my sidekick in the fight that lay ahead. Together, we would navigate the dangers of our world, and I would do everything in my power to protect him from the fate that was planned for us.

Nate's arrival began to transform me into something else—a survivalist. I couldn't passively endure this any longer. I had to protect myself, my mother, and my baby brother, Ace. Nate had isolated me from my mother, convincing her that his harsh discipline was for my own good, that he was doing what was necessary to "make a man out of me." But I could see the truth now. Nate was a plant, a GovCore agent sent to sabotage my future and keep me in check.

Nate's influence over my mother was undeniable. He had provided for us in ways we had never imagined—taking us from poverty to a semblance of middle-class stability. But those material comforts came at the cost of her autonomy, her ability to see the reality of the situation.

The arrival of Ace complicated things even further. Nate doted on him, treated him with a tenderness that was starkly absent in his interactions with me. It was as if he were two different people—one who genuinely cared for his own flesh and blood, and another who was singularly focused on his mission to destroy me.

I had to be careful. If Nate suspected that I was on to him, the consequences could be dire. So I played along, pretending to accept his authority while secretly devising ways to protect myself and my family. I started to hide my true thoughts and feelings. I put on a mask of compliance while I gathered information and planned my next move.

Every day, I became more vigilant, more aware of the subtle ways GovCore was infiltrating our lives. I began to monitor Nate as closely as he was monitoring me, watching for any slip-up, any clue that would give me an advantage. I was no longer just a child—survival had forced me to grow up fast. And I was determined to outsmart them, to find a way to break free from the grip of GovCore and the man they had sent to destroy me.

From the moment I discovered Nate's ledger, I knew I had to be careful—more careful than I'd ever been in my life. Nate was meticulous, calculating, and always seemed to be two steps ahead. But I had an advantage he didn't expect: I was watching him as closely as he was watching me.

I began to notice a pattern in his behavior. Every day, like clockwork, Nate would leave the house at the same time. He claimed he was going to work, but something about his routine felt off. On weekends, when I wasn't in school, I'd see him leave, only to return hours later, often with a strange, distant look in his eyes.

My suspicions deepened when I noticed that Nate was receiving letters in the mail—letters he quickly tucked away and never mentioned to my mother. One day, while he was out, I decided to investigate. Carefully, I retrieved one of the letters from his drawer. My hands trembled as I opened it; the fear of being caught weighed heavily on my mind.

The letter was from GovCore. As I read through the contents, the pieces of the puzzle fell into place. Nate had been indebted to GovCore ever since he was dishonorably discharged from the military due to drug and alcohol abuse. They had leveraged his desperation, forcing him into their service to pay off his debts. And now, his job was to monitor and sabotage my life, all while living under our roof.

I quickly resealed the letter with adhesive, making sure it looked untouched before placing it back where I found it. My heart raced as I realized the gravity of the situation. Nate wasn't just an abusive stepfather—he was an agent of GovCore, a man who had sold his soul to the very system that sought to destroy me.

But I couldn't let him know that I knew. I had to play along, pretending to be the obedient, fearful child he expected me to be. But in my mind, I was already devising a plan.

A few days later, I decided to follow Nate after he left the house. It was summer, and I had all the time in the world to uncover his secrets. Keeping a safe distance, I trailed behind him on my bike as he drove through the city, finally arriving at a nondescript building in a shadowy alleyway. My breath caught in my throat. I suddenly realized it was the same building I had found months earlier—the one where I'd heard the hum of machines and the murmuring voices.

Nate disappeared inside, and I knew without a doubt that he was meeting with his handlers, receiving instructions, and reporting on his progress. GovCore was in control of every aspect of our lives, and Nate was their instrument.

As the days went on, Nate's behavior became more erratic. He started drinking heavily and often passed out on the couch, oblivious to the chaos around him. One evening, as he lay in a drunken stupor, I rifled through his pockets. I was desperate to find something that might give me an edge.

What I found shocked me to my core. A small bag of crack cocaine was hidden in the depths of his jacket pocket. The sight of it made my stomach churn. I had to tell my mother. She had already endured so much, but this—this was something she couldn't ignore.

Before showing her what I had found, I waited until the next morning, when Nate had left the house. "Mom," I said, my voice trembling, "look at this."

She stared at the bag in my hand. Her face was pale and stricken. "Where did you find that?"

"In Nate's pocket," I replied. "Mom, he's not who you think he is. He's dangerous. We have to get out of here."

For the first time, I saw a glimmer of fear in her eyes. But there was something else, too—a determination I hadn't seen in years. She knew, deep down, that I was right. That this man, who had brought both comfort and terror into our lives, was a threat to our very survival.

"We have to be careful," she whispered. I could barely hear her. "We can't let him know we're planning to leave. We have to act like everything is normal."

And so, we began our preparations. Every day, while Nate was at his so-called "job," my mother and I would make phone calls, looking for a way out. We found a battered women's shelter, independently run by a kind-hearted couple, the Thompsons, who had dedicated their lives to helping women and children escape abusive situations. They agreed to take us in, no questions asked.

But the clock was ticking. In Nate's ledger, I had noticed an expiration date— a date that filled me with dread, though I didn't fully understand its significance. All I knew was that we had to get out before that date arrived, or something terrible would happen.

The tension in the house was unbearable. Nate continued to drink, oblivious to the fact that his control was slipping away. My mother played her part perfectly. She acted as though everything was fine, while secretly packing a small bag with essentials for us to take when we made our escape.

And then, the day finally came. It was a sweltering summer afternoon when Nate announced that he was leaving for work. My mother, ever the actress, gave him a passionate kiss, as if she were saying good-bye for only a few hours. "I'll see you tonight," she said. Her voice was warm and loving.

But as soon as Nate's red Thunderbird disappeared down the street, my mother turned to me. Her wide-open eyes signaled urgency. "Grab whatever you can, Toby. Just a few clothes and anything else you need. We're leaving everything else behind."

My heart raced. I hurried to my room and stuffed a small bag with whatever I could carry. My telescope, my books, my favorite toys—all of it had to stay. We couldn't afford to take anything that would slow us down.

Minutes later, the Thompsons arrived in an old station wagon. My mother, baby Ace in her arms, ushered me out of the house. We left the dogs, the television, my own room—all the things that had once brought me comfort and joy. In that moment, none of it mattered. All that mattered was getting out, getting away from Nate and the threat he posed.

As we drove away, I looked back at the house one last time. A chapter of my life was closing, a chapter filled with pain, fear, and manipulation. But ahead of us was the unknown—a future that, while uncertain, held the promise of freedom from the clutches of GovCore.

In that moment, I vowed to myself that I would never stop fighting, never stop resisting the forces that sought to control my destiny. Nate and GovCore had tried to break me, but they had only made me stronger, more determined to survive and protect my family.

We were off to a safe house, and though the road ahead would be difficult, I knew that we were finally free of Nate's grasp. But the battle was far from over. GovCore was still out there, still watching, still waiting. And I would be ready for whatever they had in store.

# CHAPTER 2:
# THE BEGINNING OF THE END

The day we escaped from Nate's grasp felt like the dawn of a new chapter, a chance to start over, to leave behind the fear and control that had dominated our lives for so long. But as we would soon learn, escaping Nate was just the beginning. The grip of GovCore reached far beyond the walls of our old house, and even though we had slipped out of their immediate grasp, their influence followed us.

Nate came home to an empty house that evening. The echo of our absence filled the rooms that had once been alive with our presence. The man who had controlled us, manipulated us, and sought to destroy us was now left with nothing but anger and confusion. I imagined him tearing through the house in a rage, searching for any clue that might tell him where we had gone. But we had been careful. We left nothing behind but the memories of our time there.

When Nate finally realized that we were gone for good, his rage probably turned cold and calculated. He would know exactly where to go. Without wasting a moment, he probably jumped into his red Thunderbird and sped off to the building hidden in the alleyway—the place where he had been receiving his orders, the place where Destinatus, the shadowy division of GovCore, operated.

Inside the dimly lit rooms of Destinatus, Nate probably reported our escape, and the news would be met with a mixture of frustration and resignation. They had lost track of us, but they knew it was only a matter of time before they would find us again. GovCore's power lay not in brute force, but in its ability to make everything seem normal, to maintain the illusion of choice and freedom while subtly guiding people toward the destinies it had planned for them.

Destinatus was more than a division of GovCore—it was the very heart of its power; it was the brain that orchestrated the complex web of control that extended into every corner of society. Often referred to as the "brains and the blade" of GovCore, Destinatus was responsible for the most sinister and calculated operations, ensuring that the corporation's grip on the population remained unchallenged.

From the moment I was born, I had been integrated into the GovCore Destiny System, though I wouldn't realize it until much later. This system was designed to track and predict every aspect of a person's life, from their potential career paths to their psychological tendencies. In the eyes of GovCore, every citizen was a potential asset—or liability—whose life had to be meticulously controlled and guided to serve the corporation's broader goals.

In GovCore's world, most people began their lives in the Collective—a broad category where the majority of the population resided. Those in the Collective lived seemingly mundane lives, working in low-level jobs with limited opportunities for advancement. This was where Nate had intended for my family and me to stay—anonymous, insignificant, and under the radar.

The Collective was characterized by its uniformity, and citizens were little more than numbers in a vast, faceless machine. Life in the Collective was marked by a sense of constant uncertainty and struggle, with GovCore ensuring that people remained dependent on the system for their survival. For most, this was their entire existence—working, struggling, and ultimately fading into obscurity.

Not everyone remained in the Collective. Some, like me, showed promise or possessed talents that caught the attention of GovCore's agents. These individuals were moved to the Ascendant Pathway where they were groomed for more prominent roles in society.

This path included careers in professional athletics, acting, or other high-profile fields that served GovCore's interests, whether through entertainment, propaganda, or by maintaining the illusion of opportunity within the system. But there was a catch—those on this track had to be easy to control or they had to offer something valuable to the GovCore elite.

Anyone who began to show signs of rebellion, or who displayed ambition that exceeded what was permitted, was quickly reassessed by Destinatus. And for those who posed a potential threat, there were other, far more dangerous paths.

I had unknowingly walked the line that led to the Disruptive Crucible Pathway, the first level of direct intervention by Destinatus. It was on this track that GovCore began to actively work against me, using Nate as their instrument.

For those like me, who began to stray from the path set by GovCore—whether by challenging the system, showing excessive ambition, or demonstrating intelligence that could pose a threat—they were placed on the Disruptive Crucible Pathway. The goal here was simple: Bring the individual back into compliance by introducing chaos and fear in their lives.

Through a series of orchestrated setbacks, Destinatus would work to entrap the individual in a cycle of short-term or long-term imprisonment, financial ruin, or social ostracism. The intent was clear—to destroy any hope of rising above the predetermined destiny and to force compliance through hardship and despair.

## THE DISRUPTIVE CRUCIBLE PATHWAY: ISOLATION AND DESTRUCTION

But I hadn't just fallen into the Disruptive Crucible Pathway; I was teetering on the edge of something far more dangerous—the Eclipsed Circuit Pathway. This was the path reserved for those who had become significant risks to the system—those who refused to conform and continued to challenge GovCore's authority.

Destinatus had already begun to lay the groundwork for my descent into this track. The horrible outcomes Ela and the other agent had planned, the isolation from friends and family, the manipulation of my mother—all of these were designed to push me into the fringes of society. On the Eclipsed Circuit, I would be systematically ostracized, labeled an outcast, and ultimately isolated from any form of support.

Those on this track often found themselves homeless, committed to asylums, or living as fugitives, constantly on the run from a system that sought to destroy them. Destinatus was adept at using traps, setups, and other forms of deception to ensure that individuals on this track were rendered powerless and alone.

## THE TERMINAL DIRECTIVE: MARKED FOR TERMINATION

And then, there was the final and most extreme measure, The Terminal Directive. This one was reserved for those who had become irredeemable threats to GovCore—individuals whose actions, influence, or knowledge could destabilize the entire system.

Once placed on this track, there was no going back. Destinatus would mark the individual for termination, and their agents would move swiftly and silently to eliminate the threat. Termination was always carried out in a way that maintained the illusion of normalcy, ensuring that society never realized the extent of GovCore's control.

## MASTERS OF DECEPTION

The true power of Destinatus lay not only in their ability to control individuals, but also in their mastery of deception. They were the unseen puppeteers, pulling the strings of society while remaining hidden in the shadows. Their ability to manipulate perception and reality was unmatched, and they excelled at placing traps—both literal and figurative—that guided individuals toward predetermined outcomes.

Destinatus used regular citizens as pawns in their game, often without those citizens ever realizing they were being manipulated. Through blackmail, bribery, or psychological manipulation, Destinatus would enlist unwitting participants to help carry out their covert operations. These people believed they were acting on their own free will, never knowing that their actions were part of a carefully orchestrated plan.

As I sat in our small apartment, listening to the sounds of the city outside, I realized that we had only just begun to escape the web that Destinatus had woven around us. Our continued struggle with poverty, and the ever-present shadow of GovCore were reminders that the fight was far from over.

But I wasn't going to let them win. My mother had fought too hard, sacrificed too much, for us to give up now. I knew that the road ahead would be difficult, that we would face more challenges, more obstacles, but I was determined to fight back. GovCore had taken so much from us, but they couldn't take everything.

This was the beginning of the end—of the life we had known, of the control GovCore had over us. But it was also the beginning of something new, a new fight, a new determination to rise above the circumstances that had been forced upon us.

We were still in the trenches, still battling against forces much larger than ourselves, but we had something they didn't expect—hope. And as long as we had that, we had a chance.

We had found temporary refuge at the battered women's shelter. The Thompsons, who ran the shelter, were kind-hearted people who provided a safe haven for those in need. For the first time in what felt like forever, we were surrounded by warmth and compassion. My mother, baby Ace, and I were welcomed with open arms, and the shelter became a small oasis of hope in the midst of our chaotic lives.

The Thompsons had helped my mother find a small apartment on the other side of town. It wasn't much—just another unit in a low-income housing project—but it was ours. It was a place where we could begin to rebuild, to create a new life away from the fear and control that Nate had brought into our lives.

But as much as we wanted to believe that we were free, the shadows of our past lingered. GovCore was still out there, watching, waiting. And they had the patience of predators. They were willing to bide their time until the moment was right to strike again.

As we settled into our new life, the reality of our situation set in. We were still living in poverty, surrounded by the same crime and despair that had defined our lives for so long. My mother, worn down by years of struggle, depended upon her two packs of cigarettes a day. In the midst of everything, that habit seemed to bring her peace.

But the cigarettes were more than a comfort—they were a ticking time bomb, planted long ago by GovCore to ensure that, even if we escaped their immediate grasp, the long-term consequences would still be devastating. The cigarettes had been reintroduced into our lives by Nate, and now, like a poison, they were taking their toll.

One afternoon, as I sat in our small apartment, my mother came home from a doctor's appointment with a look on her face that I had never seen before. It was a mixture of fear, resignation, and a deep, aching sadness.

"Toby," she said in a trembling voice, "I need to talk to you."

I looked up from the book I was reading, and my heart rate escalated. "What's wrong, Mom?"

She sat beside me, taking my hand in hers. "I went to the doctor today. They found something . . . something bad."

"What do you mean?" I asked. A knot had formed in my stomach.

She took a deep breath, and her eyes glistened with unshed tears. "They found a lump, Toby. It's breast cancer."

The words hung in the air, heavy and suffocating. I felt like the ground had been ripped out from under me, like I was falling into a void with no way to stop. Breast cancer. The words echoed in my mind, and each repetition brought with it a fresh wave of fear and despair.

"Is . . . is it bad?" I asked. My voice was a whisper.

She nodded slowly. "It's not good, Toby. The doctor said it's advanced. They're going to do everything they can, but . . . but it's going to be a fight."

The realization hit me like a freight train. My mother, the one person who had always been there for me, who had fought so hard to keep us safe, was now facing a battle that she might not win. And if she lost, what would become of us? What would happen to me and Ace?

I knew the answer. Without her, we would be lost. GovCore would have full control over our lives, and all the work we had done to escape their grasp would be for nothing. They would swoop in, take us away, and mold us into something else—whatever they wanted us to be. We would be robbed of our future—entirely stolen from us, as they had tried to do all along.

But my mother wasn't giving up. Despite the diagnosis, despite the fear, she was determined to keep fighting—for herself, for me, and for Ace. She started undergoing treatment, but it took a brutal toll. The chemotherapy sapped her strength, and left her exhausted, but she kept moving forward.

We tried to keep things as normal as possible, but normal was a distant memory. We were still trapped in poverty, still unable to afford the things that would make our lives better. I wanted to play sports, to join clubs, to do anything that would give me a sense of purpose, but every time an opportunity arose, we were faced with the harsh reality that we simply couldn't afford it.

And through it all, GovCore lurked in the background. They didn't need to send Nate after us again. They didn't need to physically force us back under their control. The cigarettes, the poverty, the cancer—it was all part of their plan, a plan that was unfolding exactly as they intended. Even when we thought we were free, we remained trapped in the web they had woven around us.

But I wasn't going to let them win. My mother had fought too hard, sacrificed too much, for us to give up now. I knew that the road ahead would be difficult, that we would face more challenges, more obstacles, but I was determined to fight back. GovCore had taken so much from us, but they couldn't take everything.

This was the beginning of the end of the life I had known, of the control GovCore had over us. But it was also the beginning of something new. A new fight, a new determination to rise above the circumstances that had been forced upon us.

As the weeks turned into months and the months into years, I watched my mother fade away. The chemotherapy had ravaged her body and spirit. Once a strong, vibrant woman who had fought to protect us, she was now only a fraction of her former self. The cancer had spread, and despite the removal of one breast, it relentlessly devoured what little strength she had left.

Ace, my little brother, was no longer a baby. Five years younger than me, he was starting to understand the world around him in ways that both amazed and terrified me. He didn't fully grasp the gravity of our situation, but he knew that something was wrong. He could see it in the way our mother moved—

slow, labored, each step a battle. He could sense it in the air, thick with the scent of medicine, and the ever-present pall of impending death.

Despite the cancer, despite the warnings from doctors and the obvious toll it was taking, my mother continued to smoke. She clung to those cigarettes like a lifeline, perhaps the only thing that felt familiar in a life that had spiraled so far out of her control. Each drag seemed to pull her further away from us, and I felt growing resentment, mingled with an overwhelming sense of helplessness.

I was sixteen now, old enough to understand the finality of what was coming, but not old enough to do anything about it. Ace was eleven, still so innocent, still so unaware of the cruel reality that awaited us. We had hospice nurses coming in and out of the house. Their presence reminded us that the end was near. They were kind, but their kindness was laced with pity, and I hated it. I hated everything about this situation—the sterile smell of the house, the hushed conversations, the way people looked at us with those eyes full of sorrow.

One morning, I woke up to the sound of the television in my mother's room. A Christian program was playing; the preacher's voice was soothing as he spoke of hope and salvation. I walked into the room, expecting to see my mother propped up on her pillows, watching as she always did. But when I looked at her, something was different. Her chest wasn't rising and falling with the familiar rhythm of her breath. Her eyes were closed, her face was peaceful in a way I hadn't seen in a long time.

"Mom?" I whispered, stepping closer to her bed. "Mom?"

There was no response. I reached out and touched her cold hand. The realization hit me, crashed over me, pulled me under. She was gone. The woman, who had given everything for us, who had fought against a system that wanted to destroy us, who had endured unimaginable pain to protect her children—was gone.

For a long time, I stood, staring at her. My mind was a whirlwind, but my thoughts made no sense. I couldn't cry. I couldn't scream. I couldn't do anything but stand there. Numb, as the reality of the change, the death, settled in.

Finally, I sat down in the chair beside her bed. I took her cold hand in mine, and started talking to her. I told her how much I loved her, how grateful I was for everything she had done for us. I told her that I would take care of Ace, that I would find a way to survive, even though I had no idea how I would do that. I kissed her on the forehead, whispered goodbye, and then I walked out of the room.

I knew what would happen next. The moment people realized she was gone, GovCore would come for Ace and me. My brother would be taken into the foster care system, and I would be thrown into whatever fate GovCore had in store for me. I couldn't let that happen. I couldn't let them win.

So, I ran. I left the house, left my mother's body lying in her bed, and I didn't look back. I knew that as soon as someone tried to reach her and she didn't answer, they would come. They would take Ace away, and they would start looking for me. But by the time they found her, I would already be gone.

## LIFE ON THE RUN

In the days that followed, I drifted from place to place, sleeping on the couches of friends who didn't know the full extent of my situation. At first, it wasn't so bad. People were kind. They let me stay for a night or two, gave me food, and offered what little they could. But kindness has its limits, and soon enough, the invitations dried up. One by one, my friends' parents started to ask questions—why wasn't I in school? Where was my mother? Why did I have nowhere to go?

When the questions became too difficult to answer, I would move on, find another couch to crash on, another friend who didn't know what was really going on. But I had no place to go. The system was closing in on me and time was running out.

I dropped out of school. There was no point in going back. Without my mother, without a home, I couldn't focus on classes. I couldn't pretend that everything was normal. Every day was a struggle to find food, to find shelter, to keep from being caught by the authorities who were undoubtedly looking for me. I had no money, no plan, no future. Just the clothes on my back and the weight of a world that seemed determined to crush me.

The hunger was the worst part. It gnawed at me, day and night, a constant reminder that I was alone, that I was losing this battle. I would go days without eating. I survived on scraps, on whatever I could find. My body grew weaker, and my mind grew more frantic as hunger ate away at my resolve. Some nights I thought about giving up, about walking into a police station, and letting them take me in. At least then I would have food, a bed, some kind of security.

But every time I thought about giving up, I remembered my mother. I remembered her lying in that bed, her body ravaged by cancer, her strength drained away. She had fought so hard to keep us safe, to give us a chance at a better life. I couldn't let her down. I couldn't let GovCore win.

So, I kept running. I kept fighting, even when it felt like there was nothing left to fight for. I moved from couch to couch until there were no couches left. I slept in parks, in alleyways, in abandoned buildings, wherever I could find a place to lay my head. I became a ghost, invisible to the world, slipping through the cracks of a society that had forgotten me.

But even as I struggled to survive, I never stopped thinking about Ace. I wondered where he was, who was taking care of him, if he was scared, if he missed me. I knew that the system had taken him in, that he was now part of the very machine that had tried to destroy our family. And I knew that if I didn't find a way to get him back, he would be lost to me forever.

Every day was a battle against despair, against the gnawing hunger, against the crushing loneliness. But through it all, one thought kept me going: I had to survive. I had to find a way to fight back, to escape GovCore's web. My mother's death was not the end—it was the beginning of a new fight, a fight for my life, for my brother's life, for the future that we had been denied.

I didn't know how I would do it, or if I could do it. But I knew one thing for certain: As long as I was breathing, I would fight. I would keep running, keep surviving, keep resisting the forces that sought to control our destiny.

Because in the end, that's all I had left—hope. And as long as I had hope, I still had a chance.

The streets became my new home, a cold and unforgiving place that cared little for a kid with no family, no money, and no future.

The desperation drove me to places I never thought I'd go. I started to get angry—angry at the world, angry at GovCore, and angry at myself for being so helpless. That anger turned into something darker, something that pushed me to do things I would never have considered before. I started hanging out with others like me—kids who had been chewed up and spit out by a system that didn't care if they lived or died. We were all lost, abandoned by the world, and we banded together out of necessity, not friendship.

We began committing small, petty crimes—nothing too serious, just enough to survive. We'd steal food from convenience stores, lift clothes from laundromats, and break into abandoned buildings for shelter. It wasn't about being criminals; it was about staying alive. I was smart, and that helped. I knew how to cover my tracks, how to stay one step ahead of the law, and how to keep us from getting caught. For a while, it worked. We were able to scrape by, just enough to keep going.

But the streets have a way of pulling you deeper into their grasp. What started as small-time theft soon escalated as our needs grew. One day, we decided to steal a car. It wasn't part of some grand plan; it was just another desperate move to get food. One of our friends worked at a fast-food restaurant and had promised to hook us up with a free meal if we could get there. So we stole a car.

My friend drove us to the restaurant where we got the food, and for a moment, things seemed normal. We were just kids, sitting in a car, eating burgers and fries, laughing like we didn't have a care in the world. But when it was time to head back, I made a decision that would change everything.

"I'll drive," I said, eager to prove myself, to feel some sense of control in a life that had spiraled so far out.

I had never driven a car before, and the nervousness in my gut should have told me to let someone else do it. But I was stubborn. I was tired of feeling powerless, tired of being scared, tired of running. I wanted to be in control, even if it was for a few minutes.

I started the car and pulled out onto the road with my hands gripping the wheel. I drove with two feet—one on the gas, the other on the brake—my nerves made me jittery, causing the car to jerk and lurch with each press on the pedals. I was all over the road, but somehow, we were making it. That is, until a police car pulled up with flashing lights.

My heart was racing; my mind was spinning. "What do I do?" I demanded. I was having a panic attack.

"Go! Just go!" my friend shouted.

I hit the gas, trying to outrun the cop, trying to escape the reality that was crashing down around me. For a moment, it seemed like we might make it. The adrenaline surged through me, bringing the false confidence that I could actually pull this off. But as we approached a stop sign, instead of hitting the brake, I slammed on the gas, and sent the car careening through the intersection and straight into a tree.

The impact was jarring; the sound of crunching metal and shattering glass filled my ears. Before I could even register what had happened, the police car pulled up, blocking us in. There was no escape. I was caught.

They arrested me on the spot, dragged me out of the wrecked car and slapped the handcuffs on my wrists. The fear and adrenaline that had driven me moments before were gone, replaced by a cold, sinking feeling in the pit of my stomach. This was it. I had been running for so long, but now they had me.

I was six months shy of eighteen, and as they booked me into the juvenile detention center, I realized how much trouble I was in. Joyriding, theft, evading the police—these were serious charges—the kind that could easily ruin what little was left of my life. But I was smart enough to know that the system didn't really care about justice, not in the way they pretended to. What they cared about was control.

I spent weeks in that detention center, waiting for them to decide my fate. Every day felt like a year; the walls closed in on me, and the weight of my decisions pressed down on my shoulders. The other kids there were hardened, angry, lost souls who had been through the same wringer I had. But I kept to myself, stayed quiet, stayed smart, and bided my time.

During those weeks, I managed to get my GED. It wasn't much, but it was something. Something to show that I wasn't just another lost cause. I still had a future, even if it was one that I couldn't see yet.

And then, one day, they came to me with an offer.

The guards led me to a small room, where two men in suits were waiting. They weren't cops, and they weren't the kind of people you saw in the detention center every day. These were men of power, of influence—the kind of men who made decisions that changed lives.

"Sit down, Toby," one of them said, motioning to a chair across from them. My mind was racing, but I did as I was told and tried to figure out what was happening.

"We've been looking at your file," the other man said, flipping through a stack of papers on the table. "You're smart. Really smart. Smarter than most of the kids we see in here."

I didn't say anything. I knew better than to speak when I didn't know what was happening.

"We have an opportunity for you," the first man continued. "You're about to turn eighteen, and with your record, you could be looking at some serious time in an adult facility. But it doesn't have to be that way."

My heart pounded in my chest as I listened to him.

The second man leaned forward. "You're a bright kid, Toby," he said. "You've got potential. The kind of potential that could be put to good use. The military is always looking for smart, capable young men like you. If you agree to enlist, we can make all of this go away. Clean your record, give you a fresh start. Four years of service, and you'll have a path to success."

It sounded too good to be true. And deep down, I knew it was. This was GovCore's doing—I could feel it. They had found me, caught me in their web, and now they were offering me a way out. But it wasn't a way out, not really. It was another way for them to control me, to use me for their own ends.

But what choice did I have? I couldn't go back to the streets. I couldn't spend years rotting away in a cell. And so, with a heavy heart and a mind full of doubts, I agreed. I signed the papers, committed to four years in the military, and sold my soul to GovCore.

*****

They released me on my eighteenth birthday, a few weeks after I signed the papers. I was free, but not really. My record was wiped clean; my crimes erased as if they'd never happened. But the cost was my freedom, my autonomy, my future. I was heading into the military, into a world that was simply an extension of the system that had been trying all my life to break me.

But there was no turning back. I was in their hands now, a pawn in their game, and all I could do was hope that I could find a way to survive, as I had on the streets. I knew they wanted to use me, to mold me into something that served their purposes. But I wasn't going to make it easy for them. I had been running all my life, but now I was done running. I was going to stand my ground, find my own way, and somehow, someway, break free from the chains that bound me.

This wasn't the end of my story. It was the beginning of a new chapter, one that would take me into a world I had never imagined, and filled with dangers I couldn't yet comprehend. But I would face them, as I had faced everything else in my life. I had to. Because as long as I was breathing, I still had a chance.

# CHAPTER 3:
## FORGE OF SHADOWS

The bus ride to the military training base felt like a journey to another world. The landscape outside the window blurred into a monotonous green and gray as we passed small towns, forests, and far-reaching barren stretches of land. Each mile we covered took me further away from the life I had known, further from the streets where I had fought to survive, and deeper into the clutches of GovCore.

The base itself was a fortress of concrete and steel, surrounded by high fences that were topped with razor wire. It seemed to be designed to keep some people in and to keep others out. The moment I stepped off the bus, the air felt different—thicker, charged with the tension of discipline and control. This was no place for a boy; it was a forge where boys were turned into soldiers, where the weak were broken and the strong were bent to the will of the system.

I had always been quick to observe, to adapt, to learn the rules of whatever game I was forced to play. This was no different, except that the stakes were higher than ever. The drill sergeants wasted no time asserting their authority, barking orders, and herding us like cattle through a series of physical and mental tests designed to strip away any sense of individuality. Here, we were not Toby, or John, or Matt—we were recruits, numbers in a file, cogs in a machine that existed to serve the interests of GovCore.

From the moment I arrived, I knew that this place was designed to break us down, to rebuild us in the image of the perfect soldier—obedient, efficient, unquestioning. But I wasn't like the others. I had been broken long before I set foot on this base. The streets had been my crucible, and I had learned how to survive in a world that was just as harsh and unforgiving as this one. I wasn't here to be broken; I was here to endure.

Days bled into weeks, and weeks into months. The training was relentless, a constant assault on the body and mind. We were pushed to our limits and then beyond, forced to find reserves of strength and resilience that we never knew we had. But while others faltered, while they succumbed to the pressure or fell in line with the superiors' expectations, I remained defiant, at lease in my heart.

I had signed the papers and agreed to serve, but that didn't mean I had surrendered. I knew that GovCore had plans for me, that they saw something in me that they wanted to control, to use for their own ends. But I was determined to resist in whatever way I could, to find a way to maintain my sense of self, my sense of purpose, even as they tried to mold me into something I wasn't.

It wasn't long before I started to see the signs—little things, things that most of the others wouldn't have noticed—the way certain officers seemed to pay more attention to me during training exercises, the way I was often pulled aside for additional tests or evaluations. They were watching me, just as they had been watching me all my life, waiting for the moment when they could reel me in, when they could finally bring me under their control.

One day, after a particularly grueling day of training, I was called into the office of one of the senior officers. His name was Major Rooke, and from the moment I walked into his office, I could tell he wasn't like the others. He had a cold, calculating look in his eyes, the kind of look that told me he was someone who played a bigger game than the rest of us were even aware of.

"Sit down, Recruit Valorian," he said; his smooth voice was authoritative. I sat down, keeping my expression neutral, my mind alert.

"You've been doing well," Major Rooke continued, leaning back in his chair as he studied me. "Better than most. You're smart, resourceful, determined. We could use someone like you in a more . . . specialized role."

I didn't respond. I knew this was coming—had known it from the moment I was handed that deal in the detention center. They hadn't brought me here to be another grunt. They had plans for me, and this was the moment when those plans would start to unfold.

"I'm offering you an opportunity," Major Rooke said; his eyes narrowed slightly as if he were trying to gauge my reaction. "A chance to be part of something bigger, something that goes beyond serving your time. You'd be given advanced training, access to classified operations. It's a fast track to a career that could take you anywhere you want to go."

It was a tempting offer, one that most recruits would have immediately jumped at without a second thought. But I knew better. I knew that everything came with a price, especially when it was offered by someone like Rooke. They weren't offering me freedom or success—they were offering me a new set of chains, a new way to control me, to use me.

"What if I say *no*?" I asked, keeping my voice steady.

Major Rooke smiled, but it was the kind of smile that didn't reach his eyes. "That's not really an option, Valorian. You've already been flagged as a high-value asset. You can either take this opportunity and make the most of it, or you can continue your training and let others make decisions for you. Either way, you're going to be serving GovCore. The choice is yours."

It was the Illusion of choice, the kind that they had been giving me all my life. But this time, I knew what they were doing. They wanted to make me feel like I had some control, like I was the one making the decision, when in reality, they had already decided my fate.

"I'll think about it," I said, knowing full well that they would be watching my every move, waiting for me to make a decision.

"You do that," Major Rooke replied. His tone suggested that he knew exactly what I was thinking. "But don't take too long. Opportunities like this don't come around often."

As I left his office, The reality of what happened soaked in. They had me where they wanted me, and now they were tightening the noose. But I wasn't going to go down without a fight. I had survived the streets, survived the system, and I would find a way to survive this too.

That night, as I lay on my thin mattress in the barracks, I thought about the path that had led me here. I thought about my mother, about Ace, about the life I had been forced to leave behind. And I promised myself that no matter what happened, I would find a way out. I would break free from GovCore's control, to live my life on my own terms.

A few days later, one early morning as the sun barely broke through the clouds, I sat alone in the mess hall. My mind was churning over the implications of what they had proposed. I knew saying no would have consequences, but I couldn't bring myself to further sell my soul.

When Major Rooke called me into his office that afternoon, I saw the expectation in his eyes. He thought he had me.

"I'm not interested," I said. My voice was steady although my heart pounded in my chest.

Rooke's expression didn't change, but the air in the room grew colder. He leaned back in his chair, fingers steepled, and simply said, "You're making a mistake, Valorian."

I could feel the unspoken threat hanging in the air as I walked out. I knew this wasn't over.

The days passed in a blur of routine and grueling training. My decision to turn down Major Rooke's offer weighed on me, but I was determined to find my own path—one that wasn't controlled by GovCore. However, they weren't about to let me slip away so easily. A few weeks after my refusal, they came back with a different offer. This time, it wasn't just about me. They dangled the one thing that could make me reconsider: Ace.

They knew how much my brother meant to me, and they played on that. "You want to see your brother, don't you? Know how he's doing? Make sure he's safe, well cared for? All we're asking is that you reconsider our offer. If you agree, we can arrange a meeting, and you can keep tabs on him while you're away. We'll make sure he's on the Ascendant Pathway. He's already showing great potential."

They had me, and they knew it. My resistance crumbled under the weight of my concern for Ace. I had spent my entire life trying to protect him, and now, the only way to ensure his safety was to play along with GovCore's plans.

So, I agreed.

They kept their promise. I was reunited with Ace in a carefully controlled meeting. The room where we met was sterile, almost clinical, but seeing Ace made it feel warmer. He'd grown since the last time I'd seen him. His wide eyes still held that familiar spark of curiosity and hope.

"Toby!" he exclaimed, running over to embrace me, and for a moment time stood still. I wrapped my arms around him, and held him tight.

"Hey, lil bro," I said, ruffling his hair. "How've you been?"

Ace pulled back slightly, and looked at me with a smile. "I'm good. The people I'm with are nice. They let me read as much as I want, and I even started learning how to play the piano!"

"That's amazing, Ace," I replied; a genuine smile crept onto my face. "I always knew you were destined for great things."

We sat down together, and I took a moment to really look at him. He seemed happy, well cared for, but I couldn't shake the guilt that gnawed at me for not being there.

"I miss Mom," Ace said as his eyes dropped to the floor.

"I miss her too," I said; my voice was thick with emotion. "But she'd be so proud of you, Ace. You're doing everything right."

Ace looked up at me with a mixture of pride and uncertainty on his face. "What about you, Toby? Are you okay?"

I nodded and forced a smile. "I'm doing what I have to, but don't worry about me. Things are going to get better, I promise. I'm taking care of some things so that when I come back, we'll be together again."

"I believe you," Ace said. His voice was full of trust that both warmed and broke my heart at the same time.

"I've got you, little brother," I said, pulling him into another hug. "Just keep doing what you're doing. I'll take care of the rest."

As we said our goodbyes, I couldn't help but feel the weight of the promise I'd just made. But seeing him safe and happy gave me the strength I needed to face what was coming. I would keep that promise—no matter what it took.

Our time together was short. Before I knew it, I was on a plane, heading to an undisclosed location near Spain. I was now part of GovCore's Dominion Division—an elite group of agents tasked with spreading GovCore's influence across the globe. My mission was clear: Help GovCore achieve world domination by bringing smaller, impoverished countries into the fold.

This wasn't just about America anymore. This was about the entire world.

When I arrived at the base of operations, I was introduced to Halix, my new partner and mentor. Halix was unlike anyone I'd ever met. From the moment I was introduced to him, it was clear that he was a man who had seen and done things that most of us couldn't even imagine. He was tall, lean, muscular and moved with fluid grace—like a predator who was always ready to strike. But it wasn't only his physical presence that set him apart; it was his mind.

Born in a small town in the eastern reaches of Europe, Halix grew up in the shadow of war. Conflict and survival were etched into his very being from a young age. He was a child of the Cold War, and his formative years were spent learning how to navigate the dangers of a world on the brink of destruction. His family was poor and lived in the remnants of a decaying Soviet-era apartment block, but Halix was different. Even as a boy, he stood out—sharp, observant, always one step ahead of those around him.

By the time he was a teenager, Halix had mastered three languages, including English, and he had an encyclopedic knowledge of history, strategy, and warfare. It wasn't long before he caught the attention of those who mattered. GovCore found him when he was seventeen and recruited him out of the ruins of his homeland with promises of power and purpose.

Halix excelled in their ranks. His intelligence was unmatched, and his ability to analyze and manipulate situations made him a valuable asset. He quickly rose through the levels of the Dominion Division specializing in psychological warfare, espionage, and covert operations. He had a reputation for being both brilliant and dangerous—a man who could outthink and outfight anyone who crossed his path.

He was the architect of some of GovCore's most successful campaigns. In the shadows of war-torn nations, Halix orchestrated coups, dismantled entire governments, and turned rebel leaders against one another, all while maintaining an air of calm detachment. He was the man they sent in when a situation required more than brute force—when precision, intellect, and subtlety were required to achieve victory.

But there was more to Halix than his reputation. Beneath the cold exterior was a man shaped by loss and hardship. His eyes, always alert, held a depth of pain that spoke of the sacrifices he had made along the way. He had lost family, friends, and even parts of himself in the service of GovCore, but he never wavered. He was a true believer, committed to the cause, convinced that the world needed to be reshaped—no matter the cost.

"GovCore saved me," he once told me during one of our late-night conversations. "They took me out of a place where I had no future, no hope. They gave me purpose, power, and a reason to keep going. I've done things that haunt me, but I've also done things that I'm proud of. In the end, it's all about survival—yours, mine, and the world's."

Halix was the perfect mentor—demanding, insightful, and unrelenting. He pushed me to my limits and tested my physical endurance as well as my mental resilience. He taught me how to think like GovCore; how to anticipate and manipulate; how to see the world as it could be, not merely as it was. Under his guidance, I became more than a soldier. I became an instrument of GovCore's will.

But despite everything he had done, despite the blood on his hands and the shadows in his past, Halix was a man of principle. He believed in what he was doing, believed that GovCore was the best hope for a world teetering on the edge of chaos. And even as I questioned my own role in their plans, I couldn't help but respect him—for his intelligence, his skill, and his unwavering dedication.

Halix was more than my trainer; he became my friend. In a world where trust was a rare commodity, we had each other's back. And even though I knew that GovCore's grip on us was tight, I felt a sense of loyalty to him that I hadn't felt in a long time. We were in this together, and I knew that whatever challenges lay ahead, we would face them side by side.

But in the end, Halix was also a reminder of the price of loyalty—to GovCore, to the mission, to the cause. He had given everything to them, and in doing so, he had lost pieces of himself that he could never get back. And as much as I admired him, I couldn't help but wonder if that was the fate that awaited me as well.

"Our job involves more than winning over governments," Halix explained one day as we reviewed our mission objectives. "It's to win over the people. We need them to believe that GovCore is their savior, that we're offering them a better life, a better future. And if they don't see it that way . . . well, we have ways of *making* them see it."

*The time passed in a blur of missions, training, and the relentless grind of our work.* Time lost its meaning as one operation bled into the next, each one was a carefully calculated move in GovCore's global chess game. Halix and I became a well-oiled machine with a partnership forged in the fires of countless missions across the globe.

We were deployed to every corner of the world, from the dense jungles of South America to the volatile deserts of the Middle East. Each mission had its own objectives—toppling regimes, securing resources, manipulating rebel factions—but the underlying goal was always the same: Extend GovCore's reach; tighten its grip on the world.

There were times when we'd find ourselves deep in enemy territory, operating behind the lines to sabotage supply chains or gather intelligence. We'd move like shadows, unseen and unheard, leaving only chaos in our wake. Then there were the diplomatic missions, where we'd dress in suits and sit across from foreign leaders, offering them deals they couldn't refuse—sweetened with exactly the right amount of coercion.

In the quieter moments, when we weren't in the field, we trained. Halix was relentless in his pursuit of perfection. He pushed us both to our limits. He taught me everything he knew, from the art of deception to the brutal efficiency of hand-to-hand combat. I learned to think like GovCore, to anticipate threats before they even appeared, to outmaneuver enemies both seen and unseen.

But with every mission, with every successful operation, I felt the weight of our mission weighing on me. The world was changing, and we were the ones changing it—one covert mission at a time. We were turning the screws on a planet that was already on the brink, tightening the noose around the necks of millions, all in the name of control.

Three years slipped away like sand through my fingers. I lost track of how many missions we'd completed, how many lives we'd upended. The man who had left America three years ago was barely recognizable now; his edges were sharpened by the constant grind of our work. His conscience was buried under layers of duty and necessity.

And then came the mission that would change everything—Operation Iron Claw. The codename itself carried a weight that I hadn't felt in years, a sense of finality that set my nerves on edge. We were being sent to a place we had been before, a place where GovCore's hold was still tenuous, where our work was unfinished.

But this mission was different. It was more than securing territory or resources—it was about making a statement, about showing the world that GovCore's will was absolute. And for the first time in years, I felt the stirrings of something I hadn't allowed myself to feel —doubt.

I could feel myself changing; the lines between right and wrong were becoming blurred. I wasn't sure anymore if what we were doing was justified. On the one hand, I saw the corruption and poverty in these countries, the way people were oppressed by their own leaders. On the other hand, I knew that GovCore wasn't exactly benevolent. They wanted control, and they would do whatever it took to get it.

But I was in too deep. I had made my choice, and now I had to live with it.

## THE MISSION: OPERATION IRON CLAW

After a few months of training and planning, it was time to put it all into action. Operation Iron Claw. Our target was a small, impoverished country that had been resisting GovCore's influence for years. The government was corrupt, but they were fiercely independent, and they had the support of several powerful rebel groups that were determined to keep GovCore out.

Our job was to change that.

We were embedded in the local embassy, working under the guise of diplomats and foreign aid workers. Our cover was perfect—no one suspected that we were anything more than well-meaning officials trying to bring aid and development to a struggling nation. But behind the scenes, we were working to dismantle the government from within, to sow discord and chaos among the rebels, to weaken their resolve and open the door for GovCore's takeover.

We used every possible tactic at our disposal. We bribed officials, blackmailed leaders, and spread propaganda that painted GovCore as the only solution to the country's problems. We even staged false flag operations—attacks that we blamed on the rebels to turn the public against them. It was dirty work, the kind that left a stain on your soul. But it was effective.

Over time, the government began to buckle under the pressure. The rebels, who were once united and strong, became fractured and divided. Their leaders turned on each other as we manipulated them from behind the scenes. The country was falling apart, and GovCore was ready to move in and take control.

But not everyone was willing to go down without a fight.

We had been in the country for months, working our way through the tangled web of alliances and betrayals that defined this place. Halix and I had spent countless hours in the shadows, manipulating leaders, sowing discord among the rebels, and slowly tightening the noose around the neck of the nation's fragile government. But this meeting was supposed to be the final stroke, the one that would break the resistance and hand the country to GovCore on a silver platter.

The room where the meeting was to take place was dimly lit, and the air was thick with the scent of burning incense—a traditional greeting, we had been told. The official we were meeting was a mid-level government minister, someone we had groomed over the past months by feeding him promises of power and protection in exchange for his cooperation. Halix had assured me that this meeting was routine, a simple exchange of information and resources, another step in our carefully crafted plan. But something felt off.

Halix noticed it too. As we walked into the room, he shot me a meaningful glance through narrowing eyes as he scanned our surroundings. There was an unnatural stillness to the place, as if nature itself was holding its breath. The official greeted us with a smile, but it didn't feel genuine.

"Gentlemen," he said, and motioned for us to sit. "Thank you for coming."

Halix spoke in a calm and measured way. "Let's keep this brief. You have the documents we requested?"

The official nodded and slid a small envelope across the table. I reached for it, but Halix's hand shot out, and stopped me. He opened the envelope himself, pulled out the papers and scanned them quickly. His face betrayed nothing, but I could see the tension in the way he held himself. I saw slight tightening of his jaw.

"These are in order," Halix said. His tone gave nothing away. He handed the papers to me, and I slipped them into my jacket pocket. "Now, let's talk about your next steps."

The official hesitated and glanced at the door. That's when I knew something was wrong.

Halix must have sensed it too. In a flash, he was on his feet and reaching for the pistol that was hidden beneath his coat. But it was too late. The door burst open, and the room flooded with armed men—rebels; their faces were hardened by years of war and hatred. They moved with the efficiency of a unit trained for ambushes. Weapons were pulled as we began to react.

"Drop your weapons!" one of the rebels shouted in a thick accent, but his voice was clear enough to convey the deadly intent behind it.

Halix didn't hesitate. He moved like lightning, aiming his gun and firing before the words had left the rebel's mouth. The first shot took the leader down. Gunfire erupted in the small room, the sound was deafening as bullets tore through the walls and furniture.

"Get down!" Halix shouted, and he pushed me behind the table as he fired at the attackers. I could hear the sharp crack of his shots. Each one was precise, and took down one rebel after another. But they kept coming, relentless, a wave of bodies intent on overwhelming us.

I scrambled to draw my weapon, but before I could, a rebel lunged at me from the side. We struggled and the room became a blur of violence and noise as I fought to keep him from driving a knife into my side. Halix was there in an instant; His gun was empty, but his instincts were razor-sharp. He grabbed the rebel by the neck, and snapped it with a quick twist before the man could react.

"We have to move!" Halix barked cutting through the chaos. He grabbed my arm and asked for my weapon, dragging me toward the window at the far end of the room.

But the rebels weren't done. More of them poured in through the doorway, and blocked our path to the exit. Halix turned to face them. His expression was grim.

"Go!" he shouted, shoving me toward the window. "I'll hold them off!"

"No!" I yelled back, refusing to leave him. But he was already on the move, holding a gun and drawing a knife from his belt and charging at the nearest rebel with a ferocity I had never seen before. He was a whirlwind of motion, shooting, slashing and stabbing; his movements were a deadly dance of survival. But I could see that it was taking everything he had. He was buying me time, sacrificing himself so that I could escape.

I hesitated for a moment, torn between staying to fight with him and following his orders to leave. But then I saw the look in his eyes—a look that told me this was it. He knew it, and I knew it too.

With a heavy heart, I turned and ran for the window. I grabbed the edge of the frame and heaved myself through, just as more gunfire erupted behind me. I hit the ground hard, rolling to absorb the impact. My heart pounded in my chest. I could hear the shouts of the rebels inside, the sound of their footsteps as they searched for me.

I didn't have time to think. I sprinted away from the building, weaved through the narrow alleys of the town, and struggled to breathe in ragged gasps. I could hear the rebels behind me. Their voices grew louder when they realized I had escaped.

The streets were a maze, but I knew them well enough by now. I ducked into a side alley, then another, moving as quickly and quietly as I could. My mind raced; adrenaline coursed through my veins. I had to get to safety, had to find a way to call for extraction. But I couldn't stop thinking about Halix, about the last moments I had seen him, fighting like a demon to give me a chance.

I finally found a small, deserted building at the edge of town. It was barely more than a shack, but it would do. I ducked inside and pulled out the small, encrypted communicator I had been given for emergencies. My hands were shaking as I keyed in the code. I prayed that the signal would get through.

"Command, this is Agent Valorian," I whispered into the device. My voice trembled. "Requesting immediate extraction. Halix is down. Repeat, Halix is down."

There was a crackle of static, and then the calm, measured voice of the operator came through. "Copy that, Valorian. Extraction team en route. Hold your position."

I slumped against the wall. My body ached from the fight and the escape. The reality of what had happened hit me hard. Halix was gone. The man who had trained me, who had been my mentor and friend, was dead. And I had left him behind.

Tears welled up in my eyes, but I fought them. There would be time to mourn later. For now, I had to survive.

The minutes felt like hours. As I waited for the extraction team, every sound outside sent a fresh wave of anxiety through me. But finally, I heard the distant thrum of helicopter blades—a sound that brought both relief and despair. They were coming for me, but I was leaving Halix behind.

The extraction team was as efficient as ever. The helicopter touched down in a clearing beyond the town, and the operatives were on the ground in seconds, moving with the precision and speed of a well-trained unit.

"Valorian!" one of them shouted and motioned for me to move. "Let's go!"

I ran toward them, my legs heavy, my heart heavier. As I reached the helicopter, I turned back for a moment and looked at the town, at the place where I had lost Halix.

"Get in!" the operative urged and pulled me into the chopper. As the door closed behind me, I knew that my life had changed forever. Halix had given everything to save me, and now I had to live with the weight of that sacrifice.

As the helicopter lifted off, and the town grew smaller beneath us, I closed my eyes, trying to block out the images that replayed in my mind—the rebels, the gunfire, the last moments of Halix's life. But the images wouldn't go away. They would never fade.

The extraction team didn't waste any time. As soon as the helicopter lifted off, the operatives inside were already coordinating the next phase. I sat in silence, staring at the floor of the chopper; my mind was numb from what had just happened. Halix was gone, but the mission wasn't over—not for them.

"Command, this is Bravo Team," one of the operatives spoke into his comms. "Primary asset secured. Secondary objective in progress. Request immediate dispatch of Recovery Unit for retrieval of Agent Halix."

The response was immediate, a cold, detached voice on the other end. "Copy that, Bravo Team. Recovery Unit en route. Stand by for further instructions."

I didn't need to ask what they meant by "Recovery Unit." I knew that they were going back for Halix, to recover his body, to bring him home. It was standard procedure, but there was nothing standard about the way it made me feel. The thought of Halix lying there, alone, while I was being flown to safety, twisted my gut with guilt and grief.

The helicopter touched down at a secure facility not far from the extraction point—a place designed for situations like this. As I was led off the chopper, my legs felt like they were made of lead, but I forced myself to keep moving. The operatives guided me inside, where I was taken to a debriefing room, stripped of my gear, and handed plain clothes.

The room was small, sterile, and devoid of warmth—a reflection of the process itself. The debriefing was thorough and methodical. They asked me to recount every detail, every action taken, every shot fired. I spoke in a monotone. The words spilled out as if they belonged to someone else. I was too numb to feel anything, too exhausted to process what had happened. All I could see in my mind was Halix, fighting alone, sacrificing himself so that I could escape.

When the debriefing was over, they moved me to another room; this one was slightly more comfortable. It was a space meant for "decompression," a euphemism for allowing agents to process the trauma of their missions. But I didn't want comfort. I wanted answers. I wanted to know if they had found Halix, if they had brought him back. But I was met with silence.

Hours passed, or maybe it was days—I couldn't tell. The hours blurred, and each moment dragged on as I waited for news. Finally, they came for me, and led me back to the helicopter. As we lifted off, I knew we were headed home, back to the States. But this time, the flight was different.

As we approached the airfield, I saw the cargo plane waiting on the tarmac. The extraction team had done their job. Halix's body had been recovered, or perhaps even negotiated back by GovCore through whatever means necessary. The details didn't matter—only the outcome. We were both coming home, but only one of us was still breathing.

The flight was long, but the landing was smooth, almost too smooth for what awaited us. As we touched down, I could see the hearse waiting at the edge of the runway. The doors of the cargo plane opened, and I watched as the team unloaded the casket, which was draped in a simple, nondescript cover. There were no flags, no honors—just the quiet efficiency of a job completed, a life ended.

I was led off the helicopter. My feet were heavy with the weight of what had happened. I couldn't bring myself to look at the passing casket. Halix was in there, the man who had saved my life, who had been more than a mentor—he had been a friend. And now, he was gone, just another casualty in GovCore's relentless march toward control.

They led me into the terminal where a car was waiting to take me back to the facility. But before I got in, I stopped and turned to face the cargo plane one last time.

"Goodbye, Halix," I whispered. The words were barely audible, even to myself. I didn't know if he could hear me, if there was anything left of him to hear. But it didn't matter. My whispered words were all I could offer.

As the car pulled away, I knew that this was only the beginning. Halix's death was a reminder of the cost of the path I was on, and the price I might have to pay before it was over. But for now, I was alive. And I would carry his memory with me, a silent vow to make sure that his sacrifice wasn't in vain.

I was a different person. The mission had changed me, hardened me in ways I hadn't expected. I had seen the worst of what humanity had to offer, and I had been part of it. I had helped to destroy a country, to break its people, to further GovCore's quest for world domination. And I had lost one of the few people I had come to trust.

# CHAPTER 4:
# RISE OF THE MACHINA

Amerca had always been a land of change, but nothing could have prepared me for what I returned to. The country I left behind had been on the cutting edge of technology, but now it seemed the future had arrived all at once. Everywhere I looked, the signs of a new order were evident—GovCore's fingerprints were on everything, and artificial intelligence was no longer just a tool; it was becoming the lifeblood of the nation.

As the plane touched down, the skyline that greeted me was almost unrecognizable. Towering skyscrapers with sleek, reflective surfaces seemed to pulse with energy; their forms blended seamlessly with the automated drones that buzzed around them like bees in a hive. Streets that once bustled with human activity were now patrolled by robotic enforcers; their movements were precise and calculated; their cold, metallic gazes scanned every passerby.

I had been gone for three years, but it felt like I had been gone for decades. The America I had known was disappearing, replaced by something colder, more controlled. GovCore had taken what was once a society of individuals and molded it into a mechanized machine, where efficiency and control were the new gods. And at the heart of it all was artificial intelligence, guiding every decision, every action, ensuring that the collective will of GovCore was enforced with unyielding precision.

As I stepped off the plane, I was met by a GovCore official—a man with a chip implanted visibly at the base of his skull; his eyes were unnervingly bright, almost too sharp. He escorted me to a waiting vehicle, and we drove in silence through the transformed city. My mind was still reeling from the sights, from the overwhelming sense that everything had changed in my absence.

When we arrived at the GovCore facility, I was taken directly to the medical wing, a place that was a cross between a hospital and a tech lab. Everything felt connected, in sync, and humming with the quiet efficiency of machines at work. I had been told they were taking Halix's body here for "processing," but nothing could have prepared me for what I was about to witness.

They had brought him back. Or at least, a version of him. I was briefed on what to expect; on the way to see him they told me that this new combination of man and technology was changing the world and giving people a second chance at life. It was helping the less fortunate achieve amazing things. Its official name was the Cerebrax Seal, or the Seal, as it had been called today. It was more than a technological marvel—it was a symbol of a new era, a dividing line between the obsolete and the enhanced.

Its inception was whispered about for years, cloaked in secrecy and speculation. GovCore had been experimenting with neural interfaces for decades and slowly refining the technology to a point where the merging of human consciousness with artificial intelligence was no longer theoretical. The Cerebrax Seal was the culmination of that research—the ultimate fusion of man and machine, capable of reshaping the very essence of humanity.

The origins of the Cerebrax Seal could be traced back to its creator, Dr. William DeCroix, a GovCore scientist who had dedicated his life to pushing the boundaries of bioengineering and artificial intelligence. DeCroix believed that the human mind was limited by its organic form, trapped in a cycle of biological decay and inefficiency. The Seal was his answer to that problem, a way to transcend the limitations of the flesh by integrating it with the limitless potential of AI.

It started small—initially developed for military purposes. The first wave of recipients were soldiers, enhanced to process information faster, heal from injuries quicker, and execute complex combat strategies with cold, machine-like precision. The early prototypes were clunky, intrusive, and prone to failure. But with each iteration, the technology improved, and soon, not only soldiers were being implanted, but also high-ranking officials, scientists, and corporate leaders. The Cerebrax Seal became the gateway to power, an enhancement that marked its wearer as part of the elite, the chosen few who had been selected by GovCore to lead the world into the future.

At its core, the Seal was a neural implant designed to interface directly with the brain, enhancing cognitive abilities, memory retention, and decision-making. But it was more than a performance boost—it was a form of control. Those who bore the Seal were plugged into GovCore's vast AI network; they were able to access information instantly, communicate telepathically with other Seal bearers, and even predict outcomes with the precision of an algorithm.

It was more than improving one's mind; it was about controlling the world around you. The Seal allowed its user to tap into GovCore's vast database of global intelligence, making its bearer capable of orchestrating events from behind the scenes, bending reality to their will. It was the ultimate tool for those who craved power and influence, and GovCore was careful to ensure that only the most loyal, the most elite, were granted access to this technology.

For those who received the Cerebrax Seal, life was a seamless blend of hyper-efficiency and omnipotence. They no longer needed to waste time on menial tasks—decisions that would take an ordinary person hours to make could now be made in milliseconds. Their cognitive abilities were augmented to the point where they could process multiple streams of information simultaneously, control entire networks with a thought, and predict human behavior with unnerving accuracy.

But this power came at a cost. The more they connected with the AI, the more they lost touch with their own humanity. Emotions became muted, replaced by the cold logic of the machine. Memories, too, became muddled as the AI stored, sorted, and optimized them for efficiency. What had once been a vibrant inner life was now reduced to a series of calculations, probabilities, and outcomes. It was as if the soul itself had been overwritten by code.

For those without the Seal, life took on a different, darker tone. The gap between the enhanced and the nonaugmented grew wider with each passing year. Those who had the Cerebrax Seal quickly rose to the top of every industry, every government, every sphere of influence. Their decisions, which were fueled by the precision of artificial intelligence, outpaced anything the average human could hope to achieve. The nonaugmented were left behind, relegated to the roles of laborers, drones, and consumers in a world where their very relevance was fading.

Society became stratified in a way that hadn't been seen since the earliest days of civilization—those *with* the Seal, and those *without*. The elite few who bore the Cerebrax Seal operated on a different plane of existence; their lives were governed by a perfect harmony of flesh and machine, while the masses struggled to keep up in a world that no longer valued their contributions.

Those without the Seal lived in a constant state of anxiety, aware that they were being outpaced by a world they no longer understood. Jobs once performed by humans were now the domain of Seal bearers or the autonomous systems they controlled. Entire industries crumbled and were replaced by GovCore's AI-driven infrastructure. And while the Seal had initially been a mark of prestige, it quickly became a necessity for survival. Those who resisted it—who clung to their humanity—found themselves at odds with a society that no longer had a place for them.

Some fought back. They formed underground resistance movements that sought to sabotage the AI networks, but their efforts were futile. The Seal bearers could anticipate every move because they were connected to

GovCore's global surveillance and predictive algorithms. To fight against GovCore was to fight against inevitability itself.

The Cerebrax Seal was doing more than changing individual lives—it was reshaping the very fabric of society. Education systems began to shift, with the brightest minds were funneled into programs designed to prepare them for Seal integration. Those who couldn't meet the stringent requirements were left behind, and their futures were limited to whatever scraps the system had left for them.

For the Seal bearers, the world was theirs to command. For everyone else, the world was slipping away.

The Seal had given GovCore ultimate control. It was more than a tool of enhancement—it was a means of ensuring that humanity itself bent to the will of those who wielded it. And in this new world, those without the Seal were no longer citizens. They were relics of a bygone era, standing on the precipice of obsolescence.

The closer we got to Halix's location, the more I could feel that something wasn't right. We arrived at a room that was secured by massive AI-infused security guards that already knew my name. "You may enter Toby Valorian," one said. I did not look directly at the things. I quickly moved past them.

Halix lay on a sleek metal table. His body was now a fusion of flesh and machine. His eyes—once so full of life and intelligence—were replaced with glowing lenses that pulsed with a faint, eerie light. His limbs, reinforced with metal and synthetic fibers, twitched occasionally as the machines around him made adjustments. But it wasn't just his physical form that had changed. I could feel it in the air, in the way the doctors and engineers whispered among themselves. Halix was no longer a man; he was something else, something . . . more.

"He's fully integrated with the AI now," one of the engineers explained in a detached, clinical voice. "His consciousness remains, but it's been enhanced with GovCore's latest algorithms. He remembers everything—his past, his skills—but now he operates at a level far beyond any human."

I stared at Halix's altered form, a mixture of awe and horror churned in my gut. This was the future GovCore envisioned—humanity augmented, controlled, enslaved by the very technology they created. But as I looked into Halix's new eyes, I couldn't help but wonder if they had taken something essential away in the process. He was alive, but he was no longer the man I had known.

"Halix," I said softly, hoping for some sign of recognition.

His head turned toward me. The lenses focused as they scanned my face. "Toby," he said. His voice was a strange blend of human warmth and mechanical precision. "It's good to see you again."

There was familiarity in his tone, but it was laced with something unsettling—coldness and efficiency that hadn't been there before. It was like talking to a ghost, a shadow of the man who had once been my friend.

"The mission isn't complete," Halix continued in his steady, emotionless voice. "Finish what we started. The mission is still critical to GovCore's plans."

Before I could respond, another figure entered the room—a woman whose presence seemed to move everything around her. The moment I saw her, something tugged at the back of my mind—a flicker of recognition, a whisper from the past. But I couldn't place it. She was beautiful, and she possessed an air of confidence that commanded the room. Her sharp, green eyes held a knowing glint as she approached me; her lips curved into a small, enigmatic smile. It was clear she knew exactly who I was, even if I didn't immediately recognize her.

"Mr. Valorian," she said and extended her hand. "It's been a long time."

I took her hand, still searching my memory for where I had seen her before. Her voice was familiar, too, a soft yet assertive tone that nagged at me. But the name she gave now—Elara Kestrel—didn't ring any bells.

"You look familiar," I admitted. I frowned slightly as I studied her face. "But I can't quite place you. Have we met before?"

Her smile widened and her face took on a hint of something playful. "Perhaps this will jog your memory. You used to know me as Ela."

And then it clicked. A wave of memories flooded back—my childhood home, the annual inspections, the day a young woman with cold eyes and a calm demeanor had come into our house and planted the seeds of chaos that would change my life forever. Ela. This was the agent who had tried to derail my future before it had even begun.

"Elara Kestrel," I repeated. The realization hit me hard. "You're the one who . . ."

"Came to your house when you were just a boy," she finished for me; her voice was silky smooth. "Yes, that was me. But I was only doing my job, Toby. I never expected to see you rise so high. You've certainly exceeded our expectations."

There was a familiarity now, a tension between us that was laced with both past bitterness and present curiosity. I was struck by how different she looked from the young woman I remembered—more refined, more poised. But the cold calculation in her eyes was still there, hidden beneath a veneer of charm.

"What do you want?" I asked, cutting straight to the point. I had learned long ago that nothing with GovCore was ever simple, and I doubted this reunion was a coincidence.

"Let's not be so formal, Toby," she replied with a hint of amusement. "We've both grown up, haven't we? I thought we could take some time to get to know each other . . . again."

The days that followed were unlike anything I had expected. Elara didn't press me for answers or demand anything from me right away. Instead, she took a different approach, one that was as unsettling as it was disarming. She invited me to accompany her to various meetings—strategic briefings, dinner events, and even a few more casual outings that seemed almost like dates. As we spent more time together, she began to reveal bits and pieces of herself—carefully, cautiously, but with an underlying warmth that suggested she wanted me to understand who she really was.

She told me about her father, Paul Kestrel, one of the founding members of GovCore. He had shaped the world as we knew it; he was a visionary who saw beyond the chaos of the present to a future where control and order were paramount. GovCore was his legacy, his attempt to impose structure on a world that teetered on the brink of destruction. It was a heavy mantle, and Elara had grown up under the weight of his expectations.

"I was never just his daughter," she confessed one evening; her voice was soft as we sat across from each other at dinner. "I was his protégé, his heir. From the moment I could walk, he was preparing me to take his place, to continue the work he started. I never really had a choice, Toby. GovCore was always going to be my life."

She was doing more than recounting her past; she was letting me in, showing me the forces that had shaped her into the woman she was now. Elara had been groomed for greatness from a young age; her education was meticulously crafted to make her the perfect leader. She was well-versed in global politics, and she had a mind for strategy that few could rival.

But what struck me most was her specialty—psychological warfare. Elara had a rare gift, an almost beautiful ability to read people, to understand what made them tick and how to use that knowledge to her advantage. She wasn't about brute force or raw power only; she was about manipulation, about bending people to her will with a subtlety that was both frightening and impressive.

Over the course of our conversations, I learned how far she had risen within GovCore. She had outmaneuvered rivals, overcome skepticism, and proved time and again that she was more than just her father's daughter. She was a force to be reckoned with, someone who had earned her place at the top of the organization through sheer determination and skill.

Yet, there was vulnerability beneath her polished exterior. A lingering shadow of doubt came from the immense pressure of living up to her father's legacy. She never said it outright, but I could sense it in the way she talked about him—in the way her voice would falter, just slightly, when she mentioned his name.

In the end, Elara Kestrel was more than the woman who had once tried to destroy me. She was a complex, multifaceted person, shaped by forces beyond her control, driven by a need to prove herself, to live up to the ideals that had been drilled into her since birth. And as much as I wanted to hate her, I couldn't help but feel grudging respect, even a strange sense of connection. She was a mirror of sorts—a reflection of what I might have become under different circumstances.

She was always impeccably dressed; she turned heads wherever we went. There was subtle flirtation in her mannerisms, the way she would smile at me across the table, or lean in a little too close when she spoke. But it was never overt, never crossed the line into something inappropriate. It was as if she were testing me, seeing how far she could push before I would push back.

Over those few days, we talked about everything and nothing. She asked me about my life, my time in the military, my thoughts on the world as it was now. I answered her questions cautiously, aware that everything I said could be used against me. But she seemed genuinely interested, even though I suspected there was always an ulterior motive.

In return, she shared bits and pieces of her own story, revealing a side of her that I hadn't expected. As for her father she repeatedly spoke of him with a mixture of pride and reverence, as if he were some sort of deity.

"He was a brilliant man," she said one evening as we sat across from each other in a dimly lit restaurant. "He saw what others couldn't—that humanity needed to be guided, molded into something greater. GovCore was his legacy, and I've spent my life trying to live up to his vision."

I was intrigued despite myself. It was clear that Elara was more than a pawn in the GovCore machine. She was a key player, someone who had been groomed from birth to take her place among the elite. And yet, there was something almost vulnerable in the way she spoke about her father, as if she were still trying to prove herself worthy of his legacy.

I chose my words carefully. "You've done well. But why are you telling me this?"

She smiled that enigmatic smile that never quite reached her eyes. "Because I want you to understand, Toby. I want you to see that what we're doing—what I'm doing—isn't only about power. It's about ensuring that the world doesn't fall apart. You've seen what's out there. You know how dangerous it can be."

The conversation drifted back to more neutral topics after that, but the underlying tension remained. I knew she was leading up to something, that all of this—the dinners, the meetings, the flirting—was building toward a request, a demand that would come when she felt the time was right.

It came three days later, during another one of our conversations. We were back at the GovCore facility, seated in a private lounge that overlooked the city's skyline. The view was breathtaking, but I couldn't shake the feeling that I was being led into a trap.

"Toby," she said. Her tone softened as she turned to face me fully. "I think it's time we talked about why I brought you here."

I nodded, finally ready to hear what I knew was coming. "Go on."

"We need you to go back," she said simply. "The mission you and Halix started—Operation Iron Claw—it's still unfinished. GovCore needs that region under control, and you're the only one who can make that happen."

She paused, letting the words sink in before continuing. "You're one of the best, Toby. You've proven that time and time again. But more than that, you have a personal stake in this. Halix . . . well, let's say he's not the same without you."

As if on cue, the door opened, and Halix stepped into the room. But this wasn't the Halix I had known. His movements were too precise, too controlled, and his eyes—those cold, artificial lenses—focused on me with an intensity that was cold and robotic.

"Toby," he said. His voice a fusion of human and computer enhancements . "It's good to see you again. I've been . . . improved since we last met. But there's still work to be done."

The sight of him, the man who had once been my mentor and friend, was now reduced to a tool of GovCore. It sent a chill down my spine. This was the future they were offering me—a life of servitude, a path that would strip me of everything that made me who I was.

"No," I said, "I'm not going back. I'm done." The words escaped my lips before I even realized I had spoken.

Elara's expression hardened, and the warmth in her eyes was replaced by a cold, calculating gaze. "You do realize what you're saying, don't you, Toby? If you walk away now, you'll be nothing. You'll return to the collective, while your brother ascends to greatness. All your accomplishments will be forgotten. Is that really what you want?"

I met her gaze, unflinchingly. "I've given you everything I can. But I'm not going to be your puppet. Not anymore."

For a moment, we stared at each other. Then she sighed, and I detected a hint of disappointment in her voice.

"Very well," she said and abruptly stood up. "If that's your choice, then so be it. But remember, Toby—once you leave, there's no coming back. You'll be another face in the crowd while the world moves on without you."

When she turned to leave, Halix followed her with the same eerie precision, but before she stepped out, she looked back at me one last time. I didn't know how to read her expression.

"Goodbye, Toby," she said. Her voice carried a finality that sent a shot of adrenaline through my body. "I hope you find what you're looking for."

And with that, she was gone, leaving me alone in the room with the consequences of my decision taking an immediate effect. Every second of silence pushed me closer to my existence as a part of the Collective. I had made my choice, but as I looked out at the city below, I wondered if I had sealed my fate.

# CHAPTER 5:
# THE NEW DAWN

Returning to the Collective was like waking up from a long, uneasy dream. After years entrenched in GovCore's secret operations, knowing their every move, their vast network of power, and their quiet manipulations, I was now thrust back into a world where I was expected to simply blend in—to pretend that I hadn't seen behind the curtain. But I had. I knew too much.

The first few weeks were the hardest. I kept my head down and sought work in the Collective like any other person my age. Work was a challenge after returning from years in the shadows of GovCore, but not in the way most people might think. With everything I had seen and learned, I wasn't merely another person searching for a mundane job—I was a weapon, sharpened by years of negotiation, deception, and covert tactics. And while I couldn't openly advertise my experiences, I knew how to use them to my advantage.

It wasn't long before I landed a job as a consultant for SpectraTech, a rising tech company that specialized in AI and digital infrastructure. They had been looking for someone with a sharp mind for negotiations, someone who could help them secure deals, navigate complex corporate environments, and identify weakness in their competitors. They didn't ask where my skills came from; they merely saw results.

In this role, I found myself sitting across from the same types of people I had once been trained to manipulate—powerful executives, government officials, corporate leaders. Negotiating had become second nature to me, and in those conference rooms, I could see the threads of influence and control woven into every decision. GovCore's shadow loomed over every deal and every handshake, but no one ever said it aloud. The corporations had learned to operate within the system, just as I had.

*The work was decent, and the pay was solid.* SpectraTech appreciated my skills, and soon I was negotiating multi-million-dollar contracts, helping them navigate the complex world of AI development. I was good at it—too good, in fact. There were moments when I could see my talents unnerving the people around me, moments when they'd look at me as if they were trying to figure out how I had gotten so good at understanding human behavior, at spotting vulnerabilities in every conversation.

But for all the success, it felt hollow. *I was still a part of the Collective.* No matter how much money I earned or how many deals I closed, I was still another piece of the machine. SpectraTech, for all its innovation and ambition, was still under the thumb of GovCore, tied to the larger system that dictated the direction of the tech world. Sure, I was living comfortably—I had an apartment, a car, a decent income—but I did not *feel* free.

Every now and then, I would catch a glimpse of my reflection in the glass-walled conference rooms. I was dressed in sharp suits, making deals that impacted the future of the tech world. But underneath it all, I was still navigating a maze that GovCore had designed. I wasn't free. Not really.

I had a knack for closing deals, for reading people and situations. But even as I climbed higher within SpectraTech, there was always a nagging sense that I was trapped. Every time I signed a new contract, every time I secured a deal, I couldn't shake the feeling that the work I was doing was still somehow controlled by the invisible hand of GovCore.

The Collective had a way of wrapping itself around you, even when you thought you'd found success. No matter how much you earned, how much you achieved, there was always a sense of limitation—a glass ceiling imposed by forces outside your control. And that was the truth I had learned through my years with GovCore. The system was more than physical control; it was about *psychological control*, and it caused you to believe you were free while they quietly pulled your strings behind the scenes.

Despite the decent pay and relative comfort, I felt like a cog in their larger machine, and my efforts ultimately fed back into a system I had once tried to resist. Even with the money I was making, I knew that true freedom was elusive.

My one solace during that time was reconnecting with my brother, Arthur, or as I had called him since we were kids—Ace. While I had been consumed by the world of GovCore, he had been carving out his own path, one that led him to the forefront of a rapidly changing technological landscape. Ace was brilliant, always had been, and now he was studying computer science at one of the top universities in the country. When we reunited, I felt a missing piece of my soul finally slotted back into place.

With the money I had saved, I was able to get a decent place to stay—nothing luxurious, but enough to feel stable for the first time in a long while, so I asked Ace to move in with me. He agreed. After Ace moved in, it seemed a part of me had been restored. I didn't realize it was missing until we were back under the same roof. For years, our lives had taken different paths, shaped by different forces. I had been consumed by the shadowy dealings of GovCore, while he had been pursuing a life built on intellect and ambition. But now, for the first time in years, we had the space to reconnect. And we needed it—both of us.

At first, it was just small talk—talk about his classes, his dorm life, and the different tech projects he was working on. Ace always had a brilliant mind, even when we were kids. Now, that brilliance was being channeled into computer science and artificial intelligence. But even with the lighthearted conversation, I could feel deeper questions lingering between us, questions neither of us had wanted to ask.

One night, after dinner, we sat down in the living room. The soft glow of the TV cast long shadows across the room. Ace turned to me with a serious expression. His voice was quieter than usual.

"Toby," he said, "I've been meaning to ask you something. About . . . Dad. Who was he? What was he like?"

The question hung in the air like a weight I hadn't realized I was carrying. I had always known that someday he would ask, but I wasn't prepared for how hard it would hit me. Ace had never known his father, which was something we had in common, and he had spent his whole life wondering, piecing together fragments from stories and memories that weren't his own.

I took a deep breath, leaned back in my chair, and tried to find words that wouldn't hurt him too much. "Your dad . . . Nate . . . was complicated. He wasn't always bad. When he first came into our lives, he seemed like a good guy. He took care of Mom, helped out around the house, and for a while, I thought he was going to be the father figure I never had."

I could see the emotion in Ace's eyes as I spoke, the longing to understand the man who had been such a mystery to him.

"But things changed," I continued in a sterner voice. "He wasn't a good person in the end. He was . . . abusive, to Mom and to me. He did things that I'll never forgive him for. But none of that was your fault, Ace. You were a kid. He loved you, in his own way, but you deserved a better father."

Ace nodded. His gaze dropped to the floor as he processed my words. I knew it wasn't easy for him to hear, but he needed to know the truth. We sat in silence for a while, letting the weight of the conversation settle between us.

Eventually, Ace spoke again in a softer voice. "I've always wondered. I mean, I knew something was off, but I never knew what really happened." His expression was a mixture of sadness and understanding. "Thanks for telling me."

After that night, something shifted between us. Walls had built up over years when we were separated. Those walls started to come down. We talked more—really talked. Not only about our past, but about everything. We stayed up late into the night, sitting at the kitchen table, talking about life, our dreams, and the future. Ace would tell me about the tech projects he was working on, and I'd share stories about my time with GovCore, carefully leaving out the darkest details, but giving him enough to understand the complexity of it all.

It felt good to have him around, to have someone who understood me in a way no one else could. We'd always been close as kids, but life had pulled us apart. Now, we were finding our way back to each other.

On the weekends, when he wasn't buried in his studies and I wasn't at work, we'd go hiking in the mountains, grab food at a local restaurant, or just hang out at home playing video games like we used to. Ace and I had always shared similar interests, and it was easy to fall back into the rhythm of being brothers again. It felt like old times, but better, because now we had the freedom to shape our own future, without anyone controlling our paths.

One night, after a long day, we sat on the small balcony of our apartment, looking out over the city lights. Ace leaned back in his chair. He looked thoughtful as he spoke." But what I want to do after I graduate . . ." he said. "There's so much happening with AI, with tech . . . I feel like the world's changing faster than we can keep up."

I nodded, knowing all too well the truth in his words. "It is. But you're ahead of it. You've got a good head on your shoulders, Ace. Whatever you do, you're going to be part of that change."

He smiled slightly, but his expression was serious. "Yeah, but I don't only want to be part of it. I want to make a real difference. I don't want to fall in line with what GovCore is doing. I've been thinking about ways we can disrupt things . . . make things better."

I was proud of the man he was becoming. He was following the rules, but he was thinking beyond them, as I always had. And I knew then that whatever was coming, we were going to face it together. Over the next few weeks I would come to discover the very thing that had brought me to the present. During one of our late-night conversations, Ace introduced me to something that would change our lives forever.

"You've got to meet this guy," Ace said one evening. His eyes lit up with excitement as we sat at the kitchen table. "He was my dorm roommate, and he's working on something revolutionary. It's going to change everything. He calls it Bitcoin."

I raised an eyebrow, intrigued but skeptical. "Bitcoin?"

Ace leaned forward. "Yeah," he said. "It's a digital currency—completely decentralized. No government, no banks, no middlemen. It's all peer-to-peer, and the best part? It's anonymous and secure. It's going to give power back to the people."

I had spent enough time within GovCore to understand the magnitude of what Ace was talking about. A currency that wasn't controlled by a centralized power? That was dangerous and disruptive. And that's exactly why it piqued my interest.

It was one of those days that felt significant from the moment it began. Ace had been talking about his roommate, Satoshi, for days, telling me how the guy was working on something that could completely upend the entire system. I didn't think much of it at first. Lots of people in tech talked big, but after the things I'd seen with GovCore, I knew real change was rare. Still, Ace was excited, and his excitement had a way of pulling me in.

When we finally met, it wasn't anything like I expected. Satoshi was quiet, almost unassuming, the kind of guy who blended into the background. He was reserved but had a quiet intensity about him, like he was always thinking ten steps ahead of everyone else. We were sitting in his dorm room. The walls were covered in code; monitors flickered with digital maps and charts. It looked more like a command center than a student's room.

"I've heard a lot about you," I said, breaking the ice. Satoshi barely looked up from his screen. He was typing rapidly, but he paused and turned to face me.

"Ace told me you'd be interested in what I'm working on," he said. His voice was calm, deliberate, like someone who had thought through every word before speaking. "Bitcoin."

I nodded. "Ace mentioned it. A decentralized currency, right? No banks, no government oversight?"

Satoshi gave a small nod. "That's right. I created it to solve one of the biggest problems in the world—control. Everything is controlled—money, information, even the value of your labor. GovCore, banks, corporations— they all dictate how wealth is distributed, how power flows. But Bitcoin . . . Bitcoin changes that. It's a currency that exists outside of the system, peer-to-peer, without needing anyone's approval."

He stood and paced as he spoke. He was clearly passionate about his creation. "It's built on something called blockchain technology—a decentralized ledger that records every transaction. Every node in the network verifies the transaction, ensuring it's legitimate, and once it's verified, it's permanently added to the blockchain. No one can alter it' no one can manipulate it. It's transparent, but it's anonymous at the same time."

I frowned, trying to grasp the concept. "Anonymous? How?"

Satoshi sat down. His eyes narrowed slightly as he leaned forward. "Every Bitcoin transaction is recorded, but the identities of the people involved are hidden behind cryptographic addresses. Instead of using your real name or a bank account number, you use a cryptographic key. People can see the transactions happening, but they can't trace them back to you—not unless you give them your key."

I glanced at Ace, then back at Satoshi. "That sounds . . . powerful. But how do you know it will work? I mean, sure, in theory, it sounds great, but has anyone actually used it?"

As Satoshi leaned back in his chair, a slight smile played on his lips. "People are already using it. In fact, it's being tested in one of the most volatile places you can think of—the dark web."

That caught my attention. "The dark web?"

"Yeah," Ace chimed in, clearly fascinated by the concept. "There's this underground marketplace called the Silk Road. People who want to buy and sell things without being traced are using Bitcoin to do it. They're avoiding banks, avoiding surveillance—GovCore can't track it. It's like this shadow economy where people are trading anonymously. Anything you can think of—drugs, contraband, weapons—it's all being bought and sold using Bitcoin."

Satoshi's smile faded as he continued. "I didn't create Bitcoin for that. I didn't create it for black market dealings. But the truth is, I can't control what people use it for. It's the price of true decentralization—once it's out there, it belongs to everyone, not just me. And that's the point."

I stared at him, trying to wrap my head around it all. "So, people are already using it to trade things the system doesn't want them to trade? And it's all invisible to GovCore."

Satoshi nodded. "Exactly. The dark web is just the beginning. Right now, people who don't want to be part of the system are finding ways to use Bitcoin. It's small—still underground—but it's growing. People are tired of being controlled. They want freedom. And Bitcoin gives them that."

There was something dangerous about what he was saying, and I knew that if GovCore ever found out how widespread Bitcoin was becoming, they'd shut it down. Hard. But at the same time, it felt like the key to everything I had been looking for. A way out. A way to truly be free.

## THE VISION FOR BITCOIN

"So, where does this go?" I asked. "What's your vision for it?"

Satoshi's eyes lit up, and his voice showed his conviction. "Bitcoin isn't just about money. It's about taking back control. Right now, GovCore, banks, and governments—they control everything. They tell you what your money is worth, how you can spend it, where you can keep it. But with Bitcoin, we take that power away from them. Imagine a world where people can trade freely, where no one can tell you how much your money is worth or take it from you. Imagine a world where people can't be tracked, where their financial lives are private."

Ace was nodding, clearly on board with the vision. But I was still wary.

"And you're not worried?" I asked, trying to probe a little deeper. "GovCore's not just going to sit back and let this happen. They thrive on control. They'll come for you, for Bitcoin, for anyone who uses it."

Satoshi paused and considered my words. "I know they'll try. But Bitcoin isn't like anything they've dealt with before. They can't seize it or shut it down. There's no central server, no company to raid, no CEO to arrest. Bitcoin is everywhere and nowhere. As long as people are using it, it exists."

I leaned back and nodded slowly. "So, it's a weapon."

Satoshi smiled. "It's a tool. But in the wrong hands, I suppose it could become a weapon."

## A NEW PATH

As the night wore on, we talked more about Bitcoin, the implications of it, and the world it could create. Satoshi's vision was clear—he wanted to disrupt the system, to give power back to the people. And while I admired that, I couldn't shake the feeling that it was going to put us in direct conflict with GovCore. They wouldn't let something like this go unnoticed forever.

Still, as I sat there, listening to Satoshi explain the intricacies of blockchain and decentralized networks, I realized that this was the key I had been looking for. A way out. A way to fight back. And with Ace at my side, it felt like we were on the brink of something monumental—something that could change everything.

The moment I understood what Satoshi was creating, I knew we were standing at the edge of something seismic. This wasn't just another tech innovation. This was a revolution—one that would uproot the very foundation of how the world worked. Bitcoin had the potential to dismantle the structures that GovCore had spent decades building. That meant it was dangerous.

During one of our late-night meetings, My mind was racing. "We need to help you get this out to the masses," I said to Satoshi; my voice was low but urgent. "It can't stay in the shadows of the dark web. If this is going to change things— if it's going to be a real challenge to the system—it has to be something everyone can use."

As Satoshi glanced up from his laptop, the glow from the screen cast sharp shadows across his face. He looked skeptical, even wary. But I could see that my words had struck a chord.

"How do we do that?" he asked quietly, and he looked back and forth between Ace and me. "This isn't just about creating a currency. It's about creating a movement."

And that's where we began. Over the next few weeks, we formed an online forum where we could start discussing Bitcoin openly with others who might be interested. The Bitcoin Meetings, as they came to be known, were held in my living room—just a few people at first. But as word spread, more and more people started showing up and packing into the small space. Some were programmers; others were tech enthusiasts, but all of them shared one thing: an intense curiosity about this new currency that seemed to defy everything they'd been told about how money and power worked.

Satoshi was the brains behind the operation, but it was clear that getting Bitcoin out into the world was going to take more than code. It was going to take people, a community, and a vision that others could rally behind. We started posting on every tech forum we could find, explaining Bitcoin, how it worked, and why it mattered. People began joining the movement, slowly at first, but then in growing numbers.

Our living room became a meeting hub, the place where minds came together to discuss and debate this new world of decentralized currency. Every night was electric with conversation. Some nights, it was only the three of us—Ace, Satoshi, and me—hunched over laptops, fleshing out ideas and building momentum. Other nights, we had tech enthusiasts crowding into the space, eager to hear more and get their hands on some Bitcoin.

But despite the excitement, one thing hung in the air: *Bitcoin had never been used on the open market.* Sure, it was being used on the dark web, but until it made its way into the regular economy, it would not be seen as legitimate by the world at large. That's when we devised the plan that would make history.

It was a simple idea—one that seemed almost laughable at the time. We'd offer someone 10,000 Bitcoin to buy two pizzas for us. It was ridiculous, almost like a joke, but the implications were enormous. If Bitcoin could be used to buy something as ordinary as pizza, then it could be used for anything. It was the first step toward breaking into the mainstream.

We posted the offer on the forum, hardly expecting to be taken seriously. After all, at that time, 10,000 Bitcoin wasn't worth much. But to our surprise, someone overseas accepted the deal. They agreed to send us two pizzas in exchange for the Bitcoin. We sent them the 10,000 Bitcoin, and a few hours later, two steaming hot pizzas arrived at my door.

We sat there, staring at the pizza for a moment, hardly able to believe it. Then we laughed—laughed until our sides hurt. It was absurd, really. But it was also profound. We had just made the first open market Bitcoin transaction in history.

We each took a slice and ate, grinning at each other like kids who had gotten away with something. But beneath the laughter, there was the understanding that *this was bigger than any of us*. Bitcoin had just proven itself, and now it was going to spread like wildfire.

With that, Ace and I became *early believers* in Bitcoin. We knew it wasn't worth much at the time, but we saw the potential. We started buying as much Bitcoin as we could from Satoshi, amassing cold wallets and digital wallets filled with thousands of Bitcoin. Each transaction felt like a gamble, but one we were sure would pay off in the long run. Satoshi owned most of the Bitcoin at that time, and while he was happy to sell it to us, I could sense his growing apprehension.

"This thing is going to change the world," I told Ace one night as we sat, staring at our computers, watching the tiny transactions of Bitcoin moving across the blockchain. "People don't realize what's coming."

Ace nodded; his excitement mirrored my own. "It's going to make us rich," he said. "Once people catch on, it's going to be huge."

And so we kept buying, hoarding our Bitcoin, waiting for the day when the world would wake up and see what we saw—that the GovCore system was on the verge of collapse, and Bitcoin was going to be one of the wrecking balls that brought it down.

# CHAPTER 6:
# THE UNSTOPPABLE SURGE

We had done it. The first Bitcoin transaction was behind us, and though it seemed like a small, almost laughable milestone at the time, it was clear that we had set something into motion—something far bigger than any of us had anticipated. The pizza transaction had opened the door, but what followed was a tidal wave of interest, innovation, and—inevitably—resistance.

## THE RISE OF DECENTRALIZED EXCHANGES

With Bitcoin now gaining momentum, people began to see its true potential. It wasn't long before decentralized exchanges—platforms that allowed people to trade Bitcoin without a middlemen or regulatory oversight—started to appear. These exchanges were revolutionary. They broke away from the traditional systems of control and allowed people from all over the world to trade Bitcoin directly with one another. No banks, no government approval—just peer-to-peer transactions happening outside the watchful eyes of GovCore.

At first, these exchanges were small, niche communities made up of tech enthusiasts, libertarians, and those who had grown tired of the suffocating control that institutions like GovCore imposed. But as more people learned about Bitcoin, the demand for decentralized trading grew rapidly. The price of Bitcoin started to rise—from mere pennies to dollars, and soon, the value shot up to fifty, then one hundred dollars per coin.

For the first time, people realized that Bitcoin wasn't just a concept. It wasn't just a tool for trading in the shadows. It was a financial revolution in the making. And with decentralized exchanges, it was growing faster than anyone could have imagined.

Ace and I watched in awe as the Bitcoin we had accumulated—thousands of coins—soared in value. What had once been worth next to nothing was now becoming a small fortune, and it was only the beginning.

## GOVCORE'S FIRST STRIKE

Of course, GovCore wasn't blind to what was happening. As Bitcoin began to spread, moving from underground communities to mainstream adoption, it quickly became a target. GovCore's first wave of combat against Bitcoin was swift and calculated. They understood the threat—this was a currency that existed outside their control, one that could undermine their entire system of financial dominance. And they weren't going to let that happen without a fight.

The first step was to attack the exchanges. GovCore tried to shut down decentralized exchanges through cyber-attacks, infiltration, and legal measures. They sent out directives labeling Bitcoin as "a threat to national security," claiming that it was being used to fund criminal enterprises, launder money, and destabilize the economy. GovCore agents began to raid the homes of those suspected of running Bitcoin operations, shutting down servers and arresting key figures.

But what GovCore didn't understand was that Bitcoin wasn't like any other system they had fought before. Decentralization was Bitcoin's strength. You couldn't take out one server or one company and kill Bitcoin. Bitcoin existed everywhere—on thousands of computers across the world, in every corner of the internet. Every time they shut down an exchange, three more popped up. And even when they thought they had crushed the community, the miners kept it alive.

## THE DECENTRALIZED POWER OF MINING

It was the miners who truly made Bitcoin unstoppable. Mining was the process by which new Bitcoins were created, and anyone with a computer could participate. All across the globe, people began mining Bitcoin—using their machines to solve complex cryptographic puzzles that verified transactions and maintained the blockchain. As more people joined the network, it grew stronger, more resilient.

GovCore failed to grasp that they weren't fighting a centralized enemy. There was no single point of failure, no CEO to arrest or corporation to dismantle. Bitcoin's power lay in its decentralization. Even if they could stop it in the United States, it was still thriving in Europe, Asia, Africa—everywhere. And every miner, every node in the network, was a part of Bitcoin's defense.

Each mining rig that came online was another crack in GovCore's armor. And as the network grew, so did Bitcoin's value. The harder GovCore fought to contain it, the more Bitcoin spread. It was like trying to extinguish a fire by throwing gasoline on it.

I watched the chaos unfold, knowing full well that this was only the beginning of the battle. GovCore wasn't used to losing, and they wouldn't give up easily. But Bitcoin wasn't something they could simply crush. It was a living organism, fueled by the very people who were tired of being controlled, tired of having their lives dictated by unseen hands.

## THE GLOBAL SPREAD

What made Bitcoin truly dangerous to GovCore wasn't just its decentralized nature—it was the fact that GovCore didn't run the entire world. They had an iron grip on the United States, but outside their borders, their influence waned. As Bitcoin spread across international lines, it became clear that it involved a global movement.

Governments in other countries—some of which had long resisted U.S. financial dominance—began to embrace Bitcoin as a way to circumvent economic sanctions and bypass international controls. In places like South America and Eastern Europe, Bitcoin was thriving, used as a tool for trade, for savings, for survival. It was no longer just a digital currency—it was a means of freedom.

As I watched this unfold, I realized that this fight was about more than money. It was about control. GovCore's control was slipping, and the more they fought back, the more people realized they didn't need to play by their rules anymore. Bitcoin was becoming a symbol of something larger—a way for ordinary people to take back their autonomy.

## THE FUTURE

The first wave of GovCore's attempts to crush Bitcoin were failing, and the stakes were raised. Ace and I continued to accumulate Bitcoin; our cold wallets grew every day, while Satoshi withdrew further into the shadows. His paranoia had reached new heights, and he became more reclusive every week. He knew, better than any of us, what was coming.

"Their first strike didn't work," Satoshi said one night. His voice was barely above a whisper. "But they'll come harder next time. GovCore won't stop. They'll find new ways to attack us. They'll come for the miners, for the exchanges . . . for us."

He wasn't wrong. Bitcoin prices were skyrocketing—drawing more attention than ever. And with attention came danger. The fight was far from over, and I knew that what lay ahead would test us in ways we couldn't yet imagine.

## THE VANISHING OF SATOSHI

One evening, Satoshi showed up at my place unannounced. He looked different—tired, and withdrawn; his eyes flickered with a sense of paranoia that had been growing for months. As he handed me a thin stack of papers, his fingers trembled slightly.

"These are the white papers," he said quietly. I took the pages, instantly recognizing the text and diagrams. They were the blueprint of Bitcoin—the system that was reshaping the financial world. But seeing the plan laid out so clearly made it seem more . . . real. More final.

Satoshi was pacing nervously across the room, "No one knows who created Bitcoin," he said. "Not in the forums, not in our group . . . we've been careful. But I think someone is starting to figure it out."

I raised an eyebrow. "What makes you say that?"

He stopped pacing and started rubbing the back of his neck. "Someone's been following me. My laptop was stolen last week. GovCore agents have been asking questions—too many questions. They're circling, Toby. I don't know how much longer I can stay under the radar."

My stomach dropped at his words. GovCore had always been a looming threat, but to hear that they were closing in on us made it all too real. I knew we had crossed a line, a line that could get us killed.

"Do you think they know who you are?" I asked.

Satoshi shook his head. "Not yet. But they're close. And that's why I've been thinking. Maybe it's time to stop. Maybe we need to pull back."

I stared at him. I was shocked. "Stop? We've come too far to stop now, Satoshi. Bitcoin's spreading. It's growing every day."

He walked over to me and sat down; his voice was quiet but insistent. "I know, but this was never supposed to be about us. That was the whole point—

Bitcoin was supposed to be decentralized. No single creator, no single leader. And yet, here we are, at the center of it."

## THE MILLION BITCOINS

Satoshi sighed deeply and looked me in the eyes. His expression was grim. "There's something else you need to know. I have . . . a million Bitcoin. Locked away in a hidden wallet. No one can access it but me."

I blinked. "A million Bitcoin? That's . . . that's insane."

"I mined it early on. Back when it was easy to mine and no one cared about it. I did it to ensure Bitcoin would have a solid foundation, but now . . . Now, I'm starting to think it's too much power for one person. I can't use it. I won't. But GovCore . . . they'll want it. And if they find out I have it, they'll try to take it."

For a long moment, we sat in silence while the weight of his words sunk in. One million Bitcoin, enough to shake the world's financial system to its core.

"So, what do we do?" I finally asked.

Satoshi leaned forward, and in a stern voice, he said, "We stop. We let Bitcoin grow on its own, without us. It's already out there. People are mining, trading . . . we don't need to be involved anymore. In fact, we can't be."

I shook my head. "You're really going to walk away from all of this?"

"It's the only way Bitcoin survives," he said firmly. "If there's a creator, if there's an owner, Bitcoin will always have a target on its back. But if we disappear . . . if I disappear . . . then Bitcoin can live on as it was meant to."

## THE DECISION TO DISBAND

We spent the next few hours discussing every angle. The risks, the dangers, what walking away would mean for all of us. By the time we reached a decision, it was clear—we would disband the group, cease our meetings, and go silent about Bitcoin in all public spaces. I could see the strain in Satoshi's face as we talked it through; the burden of what we had built weighed heavily on him. He wasn't just creating a currency. He had built something that could potentially destroy the old guard. And GovCore wouldn't let that go unanswered.

"I don't want to be found, Toby," he said finally, in a hollow voice. "I want to go back to being anonymous. That was the point all along."

I nodded as the reality settled in. "I understand. But what about Ace? What about the others in the community?"

"We all go back to our normal lives," he said firmly. "We keep the Bitcoin we have; we hold onto it, but we stop spreading the message. We let the community we've built take it from here."

There was nothing more to say. We shook hands to seal the silent agreement between us. Satoshi had given everything to create Bitcoin, but now it was time to let it go.

## SATOSHI DISAPPEARS

Days passed, and I tried to move on with life as normal. Ace continued going to school, and I focused on my work and keeping a low profile. But one day, Ace burst into the apartment with panic in his eyes.

"Toby," he said, out of breath, "Satoshi . . . he's gone. He's not coming to school anymore."

I stood quickly. "What do you mean he's gone?"

A visibly shaken Ace ran a hand through his hair. "I . . . I don't know. He left me a note. I thought he was saying goodnight or something, but now I think it was a *goodbye*."

He handed me the crumpled piece of paper. It was short, but the message was clear.

"Goodbye, friend. Take care of yourself."

Ace's voice wavered. "I tried calling him. I went to his dorm . . . nothing. He's just . . . gone."

A cold chill ran through me. Satoshi had talked about disappearing, but I didn't think it would be this sudden. And the fact that he hadn't reached out to me was even more troubling.

"We have to find him," Ace said urgently. "We can't let this go. What if GovCore . . . what if they got to him?"

## THE SEARCH BEGINS

We launched into investigation mode immediately. Every contact we had, every thread we could pull—we tugged on it. But no matter how hard we looked, there was nothing. Satoshi had vanished without a trace.

We scoured forums, trying to see if his online presence had moved elsewhere, but he had evaporated from the digital world as well. His accounts were silent. No new posts, no activity. It was like he never existed.

We even went so far as to contact some underground communities, hoping to find someone who had seen him, but they all came back with the same answer—no one knew where Satoshi Nakamoto was.

## THE UNKNOWN

Days turned into weeks, and still, there was no sign of Satoshi. The weight of his absence hung over us like a dark cloud. Ace became more anxious by the day, pacing the apartment, desperate for answers. I tried to reassure him, but the truth was, I didn't know what had happened either.

Had Satoshi disappeared on his own terms? Or had GovCore finally closed in on him? The thought gnawed at me constantly, but deep down, I thought we would probably never know.

In the meantime, Bitcoin continued to spread. But the creator—the one man who had started it all—had vanished. And in his absence, Bitcoin was no longer tethered to any one person. It had truly become decentralized, as Satoshi had intended.

But that didn't mean we were safe.

# CHAPTER 7:
# BITPOCALYPSE

I t was a strange day, one of those crisp, autumn mornings when the air felt electric. The leaves were deep orange and red, and the sky was that clear, pale blue that always comes in late October. It was October 25th—a date that would change everything for me and Ace, though we didn't know it yet.

For months, the financial landscape had been shifting beneath the surface. Bitcoin had been volatile, bouncing around between peaks and valleys. Other cryptocurrencies had started making waves too—Ethereum, a new player in the game, had risen quickly, offering something different with its ability to run smart contracts. Ethereum was like the tech behind Bitcoin's currency, but with a broader range of applications. People were starting to notice that cryptocurrencies weren't an experiment anymore. Decentralized exchanges were popping up all over the world—Japan, Europe, South America. These platforms gave people the ability to trade without the traditional banks, and GovCore had been powerless to stop it.

Bitcoin had always been volatile, shooting up and crashing down, but that day? October 25 was different. That was the day it skyrocketed to $5,000.

## THE NEWS HITS

I was at my desk, living out the normal day-to-day grind that had become our life within the Collective. Ace was at school, and I had settled into a mundane role as a consultant for a tech firm, doing what I needed to do to maintain the appearance of normalcy, of compliance. But all of that was about to change.

Ace who saw it first. He burst through the door; his eyes were wide; his face was flushed with excitement.

"Toby! Bitcoin's at 5,000!" His voice cracked with a mix of disbelief and exhilaration.

At first, I didn't react. I mean, I knew Bitcoin had been rising, but $5,000? That was insane. That kind of jump didn't just happen.

I turned to my screen, quickly pulling up the charts. And there it was. Bitcoin: $5,000.

Almost falling out of my chair. My heart pounded in my chest. I couldn't believe it. "This . . . this changes everything."

## THE CONVERSATION: REALIZATION OF WEALTH

Ace and I sat in silence for a few moments. We were staring at the screen in awe. We had invested in Bitcoin back when it was worth pennies. We had accumulated thousands—tens of thousands of Bitcoin—when no one cared about it. And now? Now we were sitting on hundreds of millions of dollars.

Ace broke the silence first. "Toby, we're . . . rich. Like, really rich. What do we do now?"

I shook my head, trying to wrap my mind around it. "I don't know, man. I mean, we always knew Bitcoin had potential, but this? This is insane."

Ace sat down across from me. His hands fidgeted. "This isn't only about the money anymore. We've been playing this game for so long—waiting, holding, believing in Bitcoin. But now . . . now it's real. This is more money than we could ever need. What do we do with it?"

I took a deep breath. "We have to be smart, Ace. This kind of wealth? It draws attention. And with GovCore already watching the crypto space, they're not going to let this slide."

He nodded. "Yeah, but . . . what about us? What does this mean for us? For our lives? We're not just regular people anymore. We can buy anything. Do anything."

## THE RISE OF ETHEREUM AND THE CRYPTO ECOSYSTEM

Ace sat up straight. His mind was racing. "And it's not just Bitcoin, Toby. Ethereum is blowing up too. People are using it to create decentralized apps, smart contracts, all kinds of things. It's like . . . the next step in crypto. We're at the center of this revolution."

I nodded and pulled up a few charts on the computer. "Ethereum has a completely different use case than Bitcoin. It's not just about currency; it's about infrastructure. People are building entire systems on top of it. It's like a new internet."

Ace grinned. "Yeah, and then there's Litecoin, Ripple, all these other cryptos starting to take off. They're all popping up on these new centralized exchanges."

We both knew what he meant. The rise of centralized exchanges had been one of the most surprising developments in the last few months. While Bitcoin and Ethereum had thrived on decentralized platforms, centralized exchanges like Coinbase and Binance had emerged, offering people an easier way to buy, sell, and trade cryptocurrencies. It was a double-edged sword.

"These centralized exchanges are making crypto more accessible," I said. "People who don't understand blockchain, who don't want to deal with cold wallets and private keys, they're using these platforms. But it's also giving GovCore an opening."

Ace frowned. "Yeah. They can regulate these exchanges, control who can buy and sell."

"Exactly. And once they realize how much money is being made, you know they're going to want to get involved."

Ace leaned back in his chair and stared at the ceiling. "So, what do we do, Toby? We're rich, but this isn't just about money anymore. This is about power. We have the ability to really break free from the system. To get out from under GovCore's thumb."

I sat quietly for a moment, considering the gravity of his words. Ace was right. We had the kind of wealth that could buy freedom—freedom from the Collective, from the Ascendant Pathway, from the entire system that GovCore had built. But that also made us a target.

"We need to be careful," I said finally. "We must protect what we have. But more than that, we have to be strategic. GovCore isn't going to let this slide. If they find out we're sitting on this much Bitcoin, they'll come after us."

Ace looked over at me. His expression was serious. "So what's the plan?"

I glanced at the computer screen again, and watched Bitcoin's value continue to tick upward. "For now? We keep quiet. We don't flaunt the money. We don't make any big moves. We stay under the radar, like we always have. But we start planning . . . planning for the future."

Ace nodded slowly, his mind clearly working through the possibilities. "And Satoshi? What do you think he's doing right now?"

I shrugged. "Who knows? He could be anywhere. But wherever he is, I'm sure he's watching. He knows what's happening."

We sat there for a long time. The weight of our newfound wealth was almost a burden. It was now more than the money. We had entered a new world—a world where cryptocurrency was changing everything, and where we were at the center of this new world.

## THE BITPOCALYPSE BEGINS

As the hours passed and we continued to talk, one thing became clear: Bitcoin's rise to $5,000 was more than a financial milestone. It was the beginning of a new era—a Bitpocalypse. The system that had been in place for so long developed cracks and was finally crumbling. Centralized exchanges were a sign that the traditional financial world was trying to adapt, but they wouldn't be able to hold on forever.

And as the price of Bitcoin continued to climb, I knew that this was only the beginning. The world was about to change, and there was no turning back.

"What do we do next?" Ace asked again. His voice was soft but full of anticipation.

I looked at him, then back at the screen. The numbers kept climbing. Each tick was higher than the last.

"We wait," I said. "For now, we wait. But when the time comes . . . we'll be ready."

The day Bitcoin hit $5,000 was the spark, but the flames of what we later called the Bitpocalypse spread faster than we could have imagined. At first, it was just us—me and Ace—celebrating the fact that we had quietly become ultra-rich and were sitting on a fortune that was now in the hundreds of millions. But as the days went on, it became clear that we weren't the only ones who were wealthy.

We started monitoring the blockchain more closely, not only for ourselves, but to see how many others were getting involved. The numbers were staggering. Millions of new wallets had popped up—many right here in America—right under GovCore's nose. People from the forums, from all walks of life, had been quietly buying and mining Bitcoin, and now they were sitting on fortunes they didn't know what to do with.

It wasn't just us anymore. Bitcoin had gone mainstream, and it was about to upend the entire financial system.

## THE FIRST SIGNS OF DISRUPTION

At first, the signs were subtle. We'd see random reports on the news—seemingly ordinary people had quit their job without notice and walked out of the office and into a brand new life. The anchors would joke about "mysterious Bitcoin millionaires," people who had somehow struck it rich overnight and decided they no longer had to play by the rules.

But then it started happening more frequently.

By mid-November, entire sections of the Collective—people who had spent their lives in mundane, repetitive jobs—began to disappear. They weren't only quitting work; they were walking away from the system entirely. Some were using their newfound wealth to buy homes, cars, whatever their hearts desired. Others were simply withdrawing from the world; they were content to live off their earnings in private. Either way, the cracks in GovCore's carefully constructed society were starting to show.

The Collective wasn't meant for people like us. It wasn't designed for millionaires, for people with freedom and autonomy. The Collective was built to control, to subdue, to keep people in their place. And now, Bitcoin was giving people an escape route—a way out that GovCore hadn't anticipated.

## THE BITPOCALYPSE CONTINUES

Ace and I watched from a distance, stayed under the radar, and observed as Bitcoin fever took hold of the world. It happening before our eyes: the Bitpocalypse.

People who had once been trapped in the system—teachers, factory workers, government clerks—suddenly had a vast sum of money. And they didn't handle it well. The first sign that things were spiraling was the sudden, extravagant displays of wealth. People were buying flashy cars, mansions, expensive jewelry—wasting their fortunes on frivolous things. It was as though the sudden wealth had intoxicated them, and they had no idea how to manage it.

Breaking news stories flooded the media. Every day there was something new: "Bitcoin Millionaire Buys Yacht, Declares He's Retired" or "Former Collective Worker Drops Half a Million on New Ferrari." The airwaves were filled with images of people who had been nobodies the day before, and were now flaunting new wealth in the most ostentatious ways possible.

But with that wealth came attention. And GovCore was watching.

## GOVCORE'S RETALIATION

The first arrests came quickly. GovCore had always prided itself on maintaining order, on keeping its citizens compliant and under control. So when people started quitting their jobs en masse, flashing their wealth, and flaunting their newfound freedom, GovCore struck back.

At first, they tried to discredit Bitcoin. News anchors and government officials appeared on TV, and warned people that Bitcoin was dangerous, unstable, a "volatile asset not recognized by the state." But it didn't matter. The more they tried to suppress Bitcoin; the more people used it. Miners kept mining; exchanges kept growing; and the price continued to rise.

Next came the arrests. GovCore agents, dressed in their slick black uniforms, began knocking on doors, dragging people out in the middle of the night. They charged them with tax evasion, financial fraud, any crime they could think of. They confiscated wealth, seized homes, and froze bank accounts.

The brutal actions were all over the news. "Bitcoin Users Face Harsh Penalties for Tax Fraud," the headlines screamed. The message was clear: If you used Bitcoin, you were a target. GovCore was not going to let this go unchecked.

But even with all their power, they couldn't stop it.

The chaos of the Bitpocalypse spread like wildfire, and with it came GovCore's inevitable retaliation. But as more people escaped the Collective and used their newfound wealth to break free from the system, GovCore was not about to sit idly by and watch their world unravel. They needed to project strength, and that's when they deployed one of their most recognizable faces: Elara Kestrel.

## THE PRESS CONFERENCE

The announcement came on a chilly afternoon in early March. GovCore had called a national press conference, and the screen flickered to life with the familiar GovCore insignia, a slick, corporate emblem that everyone in America had come to associate with control, compliance, and fear. Ace and I sat in silence, watching the broadcast unfold.

The camera panned to a sleek podium, and then Elara Kestrel appeared. She was, as usual, impeccably dressed. Her sharp features were highlighted by the cold, professional expression on her face. She walked confidently to the microphone, flanked by Halix—now in his full, cybernetic form—no longer the man I had known, but a GovCore enforcer, hardened and fused with AI technology. His cold, robotic presence gave me the chills.

Elara began to speak; her voice was calm and measured. "Ladies and gentlemen, citizens of GovCore, over the past several months, we have seen the rapid and dangerous rise of an unregulated digital currency known as Bitcoin. While it may seem like a new and exciting way to generate wealth, I am here to tell you that it is a threat to the stability of our society."

I could since Ace tense up as we listened, hanging on every word.

"Bitcoin," she continued, "is volatile, unpredictable, and most importantly, it is outside of the channels we have established to ensure the safety and security of all financial transactions. As of today, we are launching a new initiative to regulate its use. Anyone found using Bitcoin without going through the proper GovCore-authorized exchanges will be subject to arrest, investigation, and prosecution for financial fraud, tax evasion, and illicit activity."

Elara's tone remained even, but there was an underlying current of menace in her words. This wasn't a warning. It was a threat—a veiled promise of what was to come for anyone who dared defy them.

"We are also calling upon the creators of Bitcoin to come forward," she said, as her eyes narrowed. "It is time for an open and transparent discussion about what this technology is and how it can be made usable for all citizens under GovCore's governance. We encourage those responsible for this system to step forward and work with us to ensure its proper regulation."

## RECOGNIZING ELARA KESTREL

Ace turned to me. "Did she just ask for the creators to come forward?"

I stared at the screen. My heart pounded in my chest. The way she said it, the way she delivered that request—it wasn't a call for cooperation. It was a trap. They wanted to lure the creators out into the open, to capture the source of Bitcoin and tear it apart from the inside.

"They're hunting us," I muttered under my breath.

Ace looked confused. "Wait . . . what do you mean?"

I swallowed hard, trying to decide whether to tell him the truth. But then I decided to make him aware of who she was and how dangerous she and Halix were.

"That's Elara Kestrel," I said. My voice was strong and decerning.

Ace raised an eyebrow. "Okay, and . . .?"

"She's the one . . ." I continued, as my voice grew angrier, ". . . who came to our house when I was a kid. When GovCore did those inspections. She was there.

She planted that gun. She tried to ruin my life before I even knew what was going on."

I could see the realization dawning on Ace's face as the pieces fell into place. "Wait, you're saying she's been after you since then?"

I nodded slowly. "Yeah. And now she's after *us*! Only this time, it's not only about me. It's about Bitcoin. They don't want to destroy it, not yet, but they want to control it. They want to take it over from the inside, regulate it, and trap us in their system."

I took a deep breath and leaned forward as I met his gaze. "Elara Kestrel and Halix . . . they're not just names from my past. They're dangerous, Ace, more dangerous than you know."

"I ran into Elara again during my time at GovCore. Back then, she was already a high-ranking official. She was starting to climb the ranks, but even then, you could tell she was ruthless. They brought me in after a mission for what I thought was going to be a simple debriefing, but Elara was there, and she tried to seduce me. She was sharp—too sharp. Her father, Paul Kestrel, was one of the founders of GovCore, and she was born into this. She's been trained to manipulate, to control."

I leaned back and shook my head. "And Halix? He's worse. He was my mentor for a time, one of the best GovCore agents they ever had. A war veteran, an expert in combat. When they enhanced him with AI, he became more machine than man. Trust me, Ace, if he's involved, we're in real danger. He doesn't miss. He never fails."

## THE INITIATIVE TO CAPTURE BITCOIN

As we watched the rest of the press conference, Elara continued to outline the steps GovCore would take. The rhetoric was clear: Bitcoin was a threat, but it was one they were confident they could neutralize. They didn't want to appear weak, and they certainly didn't want the public to know how much damage Bitpocalypse was already causing behind the scenes.

"We will arrest anyone who defies GovCore regulations," Elara said. She was unflinching. "We're not here to crush innovation," she continued, "but to protect it. And the only way to protect it is through careful, structured oversight."

Halix stood silently beside her; his presence was a not-so-subtle reminder of the power GovCore had at its disposal. His very existence was a testament to how far they were willing to go to maintain control. Part human, part machine—he was their enforcer, their executioner.

Elara's voice softened as she delivered her final statement. A twisted smile played on her lips. "Bitcoin can thrive . . . but only under our control. We invite the creators to help us shape the future of this currency, but we will not tolerate chaos. The era of unregulated cryptocurrency is coming to an end. Compliance is mandatory."

The screen flickered and went dark, leaving Ace and me sitting in stunned silence.

## THE TRUTH REVEALED

Ace turned to me. "So, they're coming after us?" His brow was furrowed.

"They're coming after everyone," I replied, and ran a hand through my hair. "But yeah . . . if they find out we were early adopters . . . if they connect us to Satoshi, we're done. And that press conference? That wasn't an invitation. It was a trap."

Ace shook his head. His face revealed disbelief as he looked back at the screen. "And now she's calling for the creators of Bitcoin . . . for Satoshi."

"They want to control it," I said quietly. "Or destroy it, if they can't. And they're going to start hunting down anyone they think is connected."

Ace's face was pale; the importance of our situation was finally hitting him. "What are we going to do, Toby?"

I didn't have an answer, not yet. But one thing was clear: Elara Kestrel wasn't going to stop until she got what she wanted. And we were running out of time.

## A NEW THREAT

The press conference marked a turning point. It was about more than arresting Bitcoin users. GovCore was declaring war on Bitcoin—and by extension, on anyone who had been involved with its creation. And Elara Kestrel, with Halix by her side, was leading the charge.

For years, Ace and I had lived in the shadows, quietly watching as Bitcoin grew into a force capable of challenging GovCore's dominance. But now, the game had changed. GovCore was no longer just watching—they were coming for us.

And Elara Kestrel was at the helm, ready to do whatever it took to bring us down.

## ANONYMOUS AND DECENTRALIZED

Bitcoin's greatest strength was that it wasn't centralized. It wasn't tied to any one country, any one system. While GovCore controlled America, Bitcoin was thriving in Japan, Europe, South America, and beyond. Anonymous wallets allowed people to trade and hold Bitcoin without being traced. Every arrest made by GovCore only drove more people into the shadows, and more people started to adopt cryptocurrency as a way to escape the system.

We watched from a distance as the price of Bitcoin soared. It hit $10,000, then $20,000, and still, it climbed. People across the globe were becoming millionaires overnight, and with each passing day, GovCore's grip on society weakened.

Ace and I stayed lowkey, continuing to live our lives as if nothing had changed. We hadn't touched our Bitcoin in months, and we didn't plan to touch it. But we knew that the Bitpocalypse was upon us. GovCore couldn't arrest everyone. They couldn't seize every wallet. The decentralized nature of Bitcoin had made it impossible for them to control, and now, the system they had built over decades was crumbling.

We could see it happening, not just in the news, but in the streets. People were leaving the Collective, quitting their jobs, and embracing their newfound wealth. Bitcoin was everywhere, and GovCore had no idea how to contain it.

## WATCHING THE WORLD CHANGE

As the months passed, the world began to change. The old systems of control—the financial networks, the banks, the corporations—were starting to falter. Bitcoin had created a new reality, where people could break free from the chains of GovCore, and live outside the system.

But with that freedom came danger.

More people were being arrested; more wealth was being seized; and GovCore was becoming more desperate. They started passing new laws and targeting anyone who used Bitcoin. They tried to shut down the exchanges, but it didn't work. The more they fought, the more Bitcoin spread. It was like a virus—unstoppable, infecting every corner of the world.

And through it all, Ace and I remained in the shadows, watching the world unravel.

"What do you think will happen next?" Ace asked one night as we sat together, scrolling through the latest news reports.

"I don't know," I admitted. "But whatever it is, it's going to be big."

What started as a simple transaction—two pizzas for 10,000 Bitcoin—had now become a full-blown revolution. The Bitpocalypse wasn't only about wealth. It was about freedom. It was about people finally breaking free from a system that had controlled them for generations.

GovCore was losing control, and they knew it. The more they tried to stop the spread of Bitcoin, the more it grew. And now, the world was on the brink of something new, something none of us could have predicted.

But as I sat there, watching the rise of Bitcoin and the fall of GovCore's influence, I knew one thing for sure: Nothing would ever be the same again.

# CHAPTER 8:
# CONVERSION

It was time to convert our crypto into real money. The Bitpocalypse had shaken the system, but Ace and I knew that wealth in Bitcoin wasn't enough. If we wanted to live outside GovCore's reach, we needed cash, real currency we could use without raising red flags.

We decided on a plan. A vacation, we called it. But in reality, it was an operation—a journey to a country where U.S. regulation was thin, and crypto exchanges operated in the shadows. We chose Estonia, a small European nation known for its digital initiatives but also for its underworld of dubious crypto exchanges. It was perfect: out of GovCore's direct influence, and with enough tech culture to support an underground market for our needs.

## ARRIVAL IN THE SHADOWS OF TALLINN

The narrow, cobbled streets of Tallinn felt ancient. They were lined with buildings that bore the weight of centuries past. But beyond the quaint facades, a darker reality thrived. We had researched a decentralized exchange, one that posed as a sophisticated fintech company online. Its slick website boasted fast, anonymous conversions, all tax-free. However, being naturally suspicious, we knew there was a risk.

Ace and I walked into the dimly lit café that served as the front for the exchange. It was tucked in an alley, away from the bustling city center. The atmosphere was thick with the scent of stale coffee and smoke. People were hunched over laptops, quietly muttering in different languages. At the back of the room, two men in tailored suits eyed us carefully.

One of them, a broad-shouldered man with a deep scar across his cheek, nodded for us to approach. "You must be the Americans," he said in a thick, Eastern European accent. His eyes were sharp, studied us like prey with his sharp eyes.

"That's right," I replied, keeping my voice calm and steady. "We're here to convert some crypto."

His partner grinned. He was a wiry man with sunken eyes. "You've come to the right place. We handle transactions smoothly, no fuss." He gestured to the empty chairs at their table. "Sit, let's talk."

Ace and I exchanged a quick glance. We had been through enough with GovCore to know the smell of trouble, but we needed to take this risk. We sat down, each of us was ready to move at the first sign of danger.

"We're looking to convert a small amount first," I said, lying about the 'small' part. "One hundred thousand dollars in Bitcoin, to see how your process works."

The scarred man's eyebrow twitched slightly, but he masked his surprise. One hundred thousand dollars was no small amount, yet we wanted to feel them out. They exchanged a look, then nodded. "Very wise," he replied. "Send the Bitcoin to this address." He slid a piece of paper across the table. A string of alphanumeric characters were written on the paper. "Once the transaction is confirmed, we will transfer the equivalent in U.S. dollars to the account of your choosing."

## TESTING THE WATERS

Ace opened his laptop, carefully verifying the Bitcoin address. His eyes flickered with the screen's glow as he checked for signs of tampering. Satisfied, he nodded at me. I pulled out my own laptop and initiated the transfer, sending exactly one hundred thousand USD in Bitcoin. It was nerve-wracking, knowing that we were putting our money in the hands of strangers. But we were cautious; this was simply a test.

The two men watched us intently. The wiry one leaned forward. "It will take a few minutes for the transaction to confirm. We can discuss further business in the meantime. You seem like serious investors."

"We are," Ace replied curtly. "But let's get this one done first."

Minutes ticked by so slowly it felt like hours. My heart pounded as we waited. Suddenly, Ace's laptop pinged—a confirmation of the transaction. We had sent the money. Now, we watched them and waited for their move.

The scarred man glanced at his phone, and a smile slowly crept across his face. "Transaction confirmed," he said coolly. "Now, we will—"

But he didn't finish. He pocketed his phone and stood up abruptly, motioning to his partner. "Come," he ordered. "We need to . . . finalize things in the back." They moved toward a door at the rear of the café, leaving us sitting there, feeling the tension build.

"This doesn't feel right," Ace muttered under his breath. "They're stalling."

I nodded. "Let's follow them."

## THE SETUP AND SHOWDOWN

We stood and quietly made our way toward the back door. Just as I reached for the handle, the door burst open. The scarred man, now carrying a knife, lunged at us. "You Americans are too greedy!" he spat. "You think you can walk in here and take your money without risk?"

Instinct took over. I ducked, narrowly avoiding the blade, and delivered a hard punch to his midsection. He stumbled back, gasping for breath. Ace sprang into action, grabbed a chair and swung it at the wiry man who had lunged and knocked him off balance.

"Run!" I shouted, but as we turned to flee, three more men appeared from the shadows, blocking our exit.

"We know you have more," one of them hissed. He was a large brute with a jagged nose. "We'll take everything, one way or another."

Ace stepped forward. His fists were clenched, and he was ready to fight. "Over my dead body."

The room erupted into chaos. One of the men charged at me, swinging a metal pipe. I dodged, grabbed his arm and twisted it behind his back. He yelped in pain and dropped the pipe, which I quickly snatched up, and used to fend off the others.

Ace was holding his own, delivering swift kicks and punches, moving with a speed that caught our attackers off guard. But they kept coming, relentless in their pursuit of our wealth.

In the midst of the fight, the scarred man lunged at me again. This time, I wasn't fast enough. He tackled me to the ground and pinned me with his weight. "You can't escape," he growled; his knife moving dangerously close to my throat.

But Ace was there. He leaped, crashed into the man, and knocked him off me. I scrambled to my feet; adrenaline surged through my veins.

"Go! Now!" I yelled, and we bolted for the back door. We crashed through it, stumbling into a narrow alleyway. The sound of footsteps pounded behind us. They were still coming.

## THE ESCAPE

I glanced around and spotted a fire escape ladder on the side of the building. "Up there!" I pointed, and Ace didn't hesitate. We scaled the ladder and climbed onto the rooftop as our pursuers burst into the alley, shouting curses in their native tongue.

Once on the roof, we sprinted across it and leaped over the spaces between buildings—our hearts pounded in our chests. We reached the edge of the block, and slid down another fire escape, and landed in the alley below.

The sound of sirens filled the air—a police car sped down the main street with lights flashing. Our pursuers froze; they knew that any involvement with the authorities would blow their cover.

We took our chance, blended into the crowd, and slipped away as the sirens faded into the distance.

## BACK TO THE DRAWING BOARD

We didn't stop running until we reached the safety of our rented apartment on the outskirts of the city. Once inside, we locked the door and collapsed against the wall.

"That . . . was insane," Ace gasped and wiped sweat from his brow.

"No kidding," I muttered. I clutched my side where the scarred man's knife had come very close. "We're in over our heads here. But now we know . . . we can't trust anyone."

Ace nodded. His expression grim. "They tried to steal our money. And they would've killed us."

I had to catch my breath. "We need a new plan," I said. "We can't convert it all at once. We need to find smaller, safer ways. We can't risk dealing with shady characters like that again."

"Agreed," Ace replied. His eyes were steely with resolve. "From now on, we're moving in shadows."

As we sat there, our hearts still racing, we realized how close we had come to losing everything. This was the reality of holding our own fortune—no security, no help. We were our own bank, and it was up to us to protect it.

This was the beginning of the war for our wealth. And from here on out, it would only get more dangerous.

# CHAPTER 9:
## THE APEX OF FREEDOM

After narrowly escaping the black market crypto exchange in Estonia, Ace and I knew that this wasn't a game anymore. Our Bitcoin was more than digital currency—it was a target. Every step we took now had to be carefully calculated, and we had to find a safer way to convert our wealth without risking our lives.

## THE DARK WEB REVELATION

Now back in our rented apartment in Estonia, we were shaken but not defeated. It was late, and the dim glow from our laptops cast long shadows across the room. The air was thick with tension as we scoured the dark web for information. We weren't the only ones who had struck gold with Bitcoin; there were others—people who had found a way to convert their wealth into something tangible, something real.

"We need to figure out where the real rich are going," Ace muttered as his fingers moved rapidly across his keyboard. "There's got to be somewhere that's crypto-friendly. A place where people are cashing out and not lose everything."

I nodded, but kept my eyes glued to my screen. We'd heard stories, whispers of a new breed of elite—the Bitcoin millionaires—who had turned their digital fortunes into real-world riches. But where?

After hours of searching, we finally stumbled upon a series of encrypted forums in Switzerland. The name popped up repeatedly. Known for its banking secrecy and neutrality, Switzerland had become a crypto haven. We read testimonials from users who had quietly cashed out their Bitcoin at some of the most prestigious banks in the world, converting their digital currency into Swiss francs with complete discretion.

"This is it," Ace said. His voice was tinged with excitement. "We're going to Switzerland."

## SWITZERLAND: THE CRYPTO-FRIENDLY HAVEN

Within days, we were on a flight to Zurich. Switzerland's reputation for protecting privacy, paired with its growing acceptance of cryptocurrency, made it the perfect destination. As we landed, there was a palpable sense of victory in the air. The cold mountain breeze hit us as we disembarked, but inside, we felt warm. We had made it.

Our first stop was at one of the most exclusive Swiss banks: Valmont International Banking. The building was a fortress of glass and steel, perched elegantly on the edge of Lake Zurich. We walked inside where we were greeted with an air of sophistication and respect that we hadn't felt anywhere else.

A gray-haired man wearing a suit approached us immediately. "Mr. Valorian, Mr. Valorian," he said, addressing Ace and me. "We've been expecting you. Welcome to Valmont International."

We were ushered into a private meeting room where the walls were adorned with fine art. He offered drinks from a selection of expensive liquors. It felt surreal, but this was what came with immense wealth. We had become part of a different class now, a class that was treated with white-glove service and unshakable discretion.

## THE PROCESS: CONVERTING BITCOIN INTO REAL WEALTH

The bank executive explained the process of converting our Bitcoin into Swiss francs with the utmost professionalism. "At Valmont, we understand the complexities of digital assets," he said. "Your funds will be handled with complete privacy. You will be issued Swiss accounts, and we can convert your cryptocurrency at any time, at your convenience."

Ace and I shared a glance. This was exactly what we had been searching for—legitimacy, security, and the ability to finally convert a portion of our wealth into something physical, something real.

We started cautiously, transferring large amounts of Bitcoin into our new Swiss accounts. The transactions were seamless. In just a few hours, we transformed a significant portion of our digital fortune into Swiss francs. Millions of dollars were sitting safely in Swiss bank accounts, and they were ready to be converted into whatever currency we desired.

The Beginning of Extravagance

For the first time in our lives, we felt untouchable. As we walked out of Valmont International, the cold air no longer bit us. We were on top of the world. It wasn't long before we started spending our wealth, testing the limits of what our newfound fortune could bring us.

We bought a private jet—a sleek, customized Gulfstream G280—and suddenly, the world was at our feet. We flew to Monaco, where Ace picked up a neon orange Lamborghini Aventador. We stayed in the most luxurious penthouses, with sweeping ocean views, and we dined at Michelin-starred restaurants where the waitstaff knew our names before we walked in the door.

Ace was obsessed with technology, and he began purchasing every high-tech gadget he could get his hands on. Drones, virtual reality setups, military-grade encrypted phones—he was preparing for something, though at the time, I didn't understand what.

We flew to Africa and bought land—a sprawling estate on the coastline of Mozambique, untouched by tourism. We purchased a private island in the South Pacific, where we started building a futuristic retreat—our sanctuary, far from the prying eyes of GovCore.

The more we spent, the more invincible we felt. But we sensed a threatening undercurrent, and it began to bubble beneath the surface.

## THE UNDERPINNINGS OF GOVCORE'S ANGER

GovCore watched silently while Ace and I lived in luxury. At first, the government had been indifferent, content to dismiss Bitcoin as a passing trend. But as more people began to quit their jobs, buy cars, homes, and live outside the system, GovCore's interest sharpened.

Reports of arrests started coming in—not for Bitcoin itself, but for things like tax evasion and fraud. The government needed a way to clamp down, and they were beginning to find it. The pathways of those who had quit their jobs to live off Bitcoin wealth were quietly being shifted. People were being pushed from the Collective into the Exiled, where they were deemed a danger to the system. They were losing their freedom—without realizing it.

Meanwhile, we were too busy living life to see it coming.

We attended parties with European elites, rubbed shoulders with tech billionaires and royalty. Everywhere we went, people were fascinated by Bitcoin, with the revolution it was creating. They wanted in. But what we didn't realize was that with every person we talked to, with every new convert, we were slowly inching closer to GovCore's wrath.

## OBLIVIOUS TO THE COMING STORM

In those days, Ace and I didn't feel the danger lurking beneath the surface. We were blinded by our wealth, by the thrill of living outside the system that had oppressed us for so long.

One night, after a long day of flying across Europe in our jet, we sat on the terrace of our Swiss villa, sipping the finest wine money could buy.

"You know," Ace said, gazing out at the mountains, "we've done it, man. We've finally broken free. We're the richest we've ever been, and there's no going back."

I raised my glass in agreement and smiled as the warmth of the wine spread through me. "Yeah. We're untouchable."

But as we sat there, enjoying the fruits of our labor, GovCore was already laying the groundwork for their next move. We had no idea that the freedom we had fought for, the wealth we had amassed, was about to make us targets in ways we couldn't even imagine.

The Bitpocalypse had changed everything—and now, the shadow of retaliation loomed just ahead of us.

## THE RETURN

After months of living the high life, dating beautiful women, wearing tailor-made suits, and flying in private jets to exotic destinations, Ace and I began to feel the tug of reality. We had been traveling, carefree, for almost a year. The jet-setting lifestyle had swallowed us whole, and we reveled in it. Private villas in the South of France, land in Mozambique, penthouses in Dubai—we were living a dream that few could imagine.

But, as we sat in yet another opulent suite, surrounded by luxury, it hit us.

"Man," Ace started, swirling a glass of whiskey in his hand. Suddenly he grew distant. "We can't stay out here forever."

I looked over at him. My mind was starting to catch up with reality. "What do you mean?"

"We've been gone for almost a year, Toby. No jobs, no school. GovCore's probably kicked me out, and you . . . well, I don't even know what they've done to your position by now," Ace said, running a hand through his hair. "We've been so caught up, man. We're US citizens, and sooner or later, we've got to return."

I leaned back, thinking about the truth of what he said. He was right. The world back home had moved on without us. "You think they've noticed?"

"Of course they've noticed." Ace took a deep breath. "You were probably fired months ago, and I bet my enrollment's been canceled. We're ghosts now, Toby. Everything we've built back home is . . . gone."

I dropped my head and frowned. "Yeah . . . we're still rich though. We've got more money than we know what to do with. But we can't exactly flaunt it back in the States. Not with GovCore watching."

"That's exactly it," Ace replied. "We've got everything we could ever want, but how do we go back and act like nothing's changed? How do we stay under the radar?"

We sat in silence for a moment. The world we had left behind was calling, but we couldn't just waltz back into it. Not after the Bitpocalypse. Not after all the wealth we'd accumulated, and definitely not with GovCore's eyes scanning for anyone who might have ties to Bitcoin. It had grown into something massive—an unstoppable force—and now GovCore was watching, hunting for the creators, tightening its grip on anyone who seemed too . . . free.

"We'll have to play it smart," I said finally. "Keep our assets quiet. We'll go back, act like nothing's happened, maybe buy a small house and live low-key. No more flashy stuff. We're going to need a normal cover, at least for now."

Ace nodded. "Yeah, you're right. Things have to go back to way they were before left. We'll need to move back like we're average citizens, nothing to raise red flags. We've got the apartment still, right?"

"Yeah," I confirmed. "That's all we have left. Everything else . . . well, they've probably stripped us from the system. We'll have to use cash and transfer funds slowly. No Bitcoin accounts in America—that's our safeguard. But we've got to reestablish ourselves as normal people."

Ace sighed as he looked out at the setting sun over the horizon. "We'll need to buy a house, something modest. Stay off the radar. Maybe open a few small businesses, reinvest, but stay quiet about the money. Blend in."

My mind raced with plans, but there was still that nagging feeling. GovCore wasn't going to let this go. I could feel their presence growing, creeping around in the shadows. We weren't the only ones who had cashed in on Bitcoin, and every day more people were quitting their jobs, flaunting their wealth. The system was breaking, and GovCore was preparing to retaliate.

"I've been keeping an eye on the news," I said, pulling out my phone. "GovCore's clamping down hard, man. People are getting locked up for the smallest infractions. They're shifting people from the Collective into the Exiled pathway, quietly. It's not even being broadcasted anymore."

Ace nodded; his expression dark. "Yeah, I've seen that. They're arresting people. Making them disappear. Bitcoin's hit them harder than we thought. And now they're going after anyone connected to it."

"We'll be next if we're not careful," I said. "We need to get back, keep our heads down, and make sure no one knows how deep we are into this."

Ace's eyes flickered with determination. "We've got this. We'll get back, stay quiet, and figure out what to do next. But this . . . this is bigger than we thought."

## THE DEPARTURE: COMING HOME CHANGED

A week later, we were on our private jet, flying back to the US. The plane glided silently beneath us; the sky stretched endlessly outside the window. Ace and I sat in silence for a while, both deep in thought. Everything had changed, and yet, nothing had changed. We were returning to America, but we weren't the same people who had left nearly a year earlier.

Ace finally broke the silence. "We've been gone so long," he said. "It doesn't even feel real, man. It's like the whole world moved forward, and we're just coming back from a dream."

I nodded. "Yeah. A year. It went by fast, but everything is different now."

Ace shifted in his seat. He was staring out the window. Finally, he asked: "Do you think we can pull it off? Live normal lives? Stay off the radar?"

I thought about it for a moment. "We have to. We don't have a choice. We play it smart, stay under the radar, and no one will ever know how much we have."

Ace smiled slightly. "Yeah, we're rich, man. We've got everything. Houses, boats, property. We could buy a small country if we wanted to."

"And that's the problem," I replied, chuckling softly. "We have to make sure no one knows just how much we have. But for now, let's enjoy the ride. We've earned it."

As the jet glided smoothly through the clouds, I couldn't shake the feeling that we were flying straight into a storm. GovCore was tightening its grip, and it was only a matter of time before they started coming after people like us. But for now, we had each other, our wealth, and a plan.

The world had changed. So had we.

And as we touched down in the U.S., we knew that things would never be the same again.

# CHAPTER 10:
# THE CALM BEFORE THE STORM

When we landed back in the States, everything seemed eerily normal. The flight had been smooth, the skies clear. As we stepped off the jet, there were no red flags that we noticed, no officials waiting for us, no one paying particular attention. But we were wrong. We'd been gone nearly a year, jet-setting across Europe, Africa, and the world, and in that time, the world had changed.

But we didn't know that yet.

Ace and I were relieved to get back to our apartment. Everything was just as we left it. The place had gathered dust, but nothing had been disturbed. We felt we could pick up our lives where we left off. I remember standing in the doorway, gazing around, almost nostalgic for the life we used to have. Back before we realized how much power and wealth we were sitting on.

"Nothing's changed," Ace muttered as he dropped his bag by the couch and looked around. "It's like we never left."

I nodded, stepped inside, kicked off my shoes. "Yeah . . . feels strange, doesn't it? Like time's been frozen in here."

"Frozen, maybe. But the world's moving faster than ever," Ace replied, flipping on the TV. News channels flickered across the screen; their flashy graphics and bold headlines assaulted us with everything we had missed.

The first thing that caught my attention was a press conference. The sleek and polished figure of Elara Kestrel stood front and center, her cold eyes scanning the room of reporters like she was hunting prey. Behind her stood Halix, as silent as ever, but there was something different about him. The air around him was tense, and his stance was robotic. GovCore was pushing its agenda harder now.

"We're offering one million dollars," Elara said in a voice that was steady and authoritative. "We're offering one million dollars to anyone who comes forward with information about the creators of Bitcoin. If you have leads, knowledge of who might have developed this currency, we urge you to cooperate. This is a matter of financial integrity but also of national security. Bitcoin has destabilized the economy, and GovCore needs to restore balance."

Ace sat up straight, his eyes opened wide as he glanced at me. "Wait . . . what?"

I stared at the screen. My stomach churning. "They're putting a bounty out on us."

"One million dollars . . ." Ace whispered, rubbing his temples. "They're looking for us, man. They're looking for Satoshi, for anyone connected to Bitcoin. This is worse than we thought."

The broadcast continued, showing clips of GovCore agents searching homes, seizing Bitcoin and property from people who had been accused of using the currency illegally. Elara's words rang in my head. Bitcoin has destabilized the economy. This wasn't about money anymore—it was about control. GovCore needed to squash the uprising, to regain the grip they were rapidly losing.

"They're making it sound like a public service," I said, shaking my head. "But they're just tightening their stranglehold. Offering rewards, making people turn on each other. GovCore doesn't care about stability—they care about power."

Ace was pacing now, and anxiety was visible in every movement. "What are we going to do, Toby? We can't hide forever. People are going to start looking, and we just came back in a private jet. That's gonna get noticed."

I rubbed my face, trying to think. "We need to lay low. We still have our apartment. We haven't spent anything crazy here yet. If we stay quiet, don't draw attention, we can figure out our next move."

But even as I said the words, I knew it wasn't that simple. GovCore was ramping up their efforts, sending out their agents in force. They were hunting for anyone tied to Bitcoin, and with the promise of a million-dollar reward, people would be more than willing to sell out their neighbors, their friends. We weren't safe.

I changed the channel, trying to get my mind off the looming threat, but no matter where I turned, the same theme echoed across the news: GovCore was coming for Bitcoin. They were seizing assets, interrogating people, offering incentives to anyone who could turn in the creators of the currency. We watched people get dragged from their homes, some in tears, others in disbelief, as GovCore agents sifted through their belongings.

"It's getting worse," Ace muttered, collapsing into a chair. "We thought we had all this time. We thought we were safe . . . but now, they're right on our heels."

I sat down across from him trying to think, but my mind was racing. "We need to stay ahead of this. The inspections are coming soon."

"I almost forgot about the annual inspections," Ace said. His voice was filled with dread. "I bet they've beefed up agents this time!"

I nodded grimly. "They're looking for Bitcoin. They've started asking questions, going through people's computers, detaining folks. If they find anything on us . . . we're done."

Silence fell between us as we let it all settle in. We had been living the high life, free from the system; but now, the system was coming for us. And we weren't ready. I glanced at the TV again, watching as Halix, now a high-ranking GovCore enforcer, stood beside Elara in the press conference. He was more machine than man now, his body stiff, his expression unreadable.

"I know Halix," I said, breaking the silence. "He's dangerous, Ace. He won't stop until he gets what he's after."

"Then what do we do?" Ace asked; his voice was barely above a whisper. "How do we stop them from finding us?"

I looked him in the eye. "We play their game, for now. Keep our heads down; stay under the radar, and when the time comes, we'll be ready."

But even as I said the words, the undercurrent of the storm lingered. The walls were closing in. GovCore was tightening its grip, and the inspections were only days away.

Something bad was coming. We could feel it in the air. But for now, all we could do was wait—and hope we had enough time to escape the noose that was tightening around us.

As the news continued to flash reports of seizures, arrests, and crackdowns, Ace and I sat in silence, the tension thickening. We had enjoyed our freedom, but now, the real game was beginning, and GovCore was about to make their next move. The annual inspection was around the corner, and we were running out of time.

The Interrogation

The day finally arrived for the inspection, and we knew it would be unlike any other before. We sat in silence then... Suddenly there was a knock on the door sharp, a sound that sliced through the air like a blade. Ace and I exchanged a glance, both of us frozen for a moment. We knew this was coming, but knowing it was coming didn't make it any easier.

"It's them," Ace muttered under his breath.

I took a deep breath, nodded, and walked to the door. As I pulled it open, my suspicions were confirmed. Halix stood there, looking larger than life in the doorway, his dark coat billowing slightly in the breeze. His eyes, now cold and metallic, settled on me.

"Toby," he said; his voice was calm, but it carried tone that made the hair on my arms stand up. "It's been a while."

Behind him stood a team of GovCore agents, all equipped with the latest tech, scanning the building with devices I couldn't begin to comprehend. They looked more like machines than people, a reminder of how much GovCore had integrated its AI-fused soldiers into society.

I tried to keep my voice steady, playing it cool. "Yeah, Halix. It has. What brings you to my humble abode?"

Halix didn't blink. His expression remained unreadable, though I could tell by the way his eyes studied me that he hadn't forgotten anything about our past. "Annual inspection," he said simply, stepping past me into the apartment without waiting for an invitation.

Ace shot me a glance as the rest of the GovCore team filed in behind Halix. They moved with purpose, like they'd done this a hundred times before— because they had. We tried to stay calm, playing the part of two average citizens just trying to get by.

"Nice place," Halix remarked, glancing around as if he were genuinely interested. His gaze landed on Ace for a moment before shifting back to me. "I hear you've been . . . off the radar for a while. Not going to work. Author out of school. Care to explain?"

I shrugged, keeping my voice casual. "We decided to take some time off. I mean, we've been working hard, studying hard. Thought we deserved a break."

Halix smirked, but it was a cold, detached sort of smile. "Funny, Toby. That's not what your pathway indicates. You know how it works—you're supposed to be contributing. Staying in your lane. But now . . ." He paused, letting the words hang in the air, "... it looks like you're veering off track."

I crossed my arms and leaned back against the wall. "What's that supposed to mean? We haven't done anything wrong."

Ace shifted uncomfortably beside me, sensing the tension rising. He knew I was starting to push buttons, but I couldn't help it. Halix's presence brought out a rebellious streak in me. I had to fight the urge to lash out.

Halix took a step closer, his eyes narrowed "It means, Toby, that you've been acting suspicious. You disappear for nearly a year, show up unannounced, and neither of you has been going to your designated work or school. Not exactly normal behavior, is it?"

Ace cleared his throat, trying to ease the situation. "Look, we were just traveling. You know, seeing the world. It's not a crime, is it?"

"Not a crime," Halix replied, still focused on me, "but it's enough to raise questions. And questions need answers. That's what I'm here for."

One of the agents moved to our living room and set up his scanner in front of our computer. A low buzz filled the air as the device started going through all our data. Another agent began rifling through our belongings, scanning everything from our books to our kitchen utensils. It was invasive, but I knew it was all part of the game.

"You're not going to find anything here," I said. I met Halix's gaze head-on. "We've got nothing to hide."

He ignored my comment. "Bitcoin," he said. The word hung between us like a loaded gun. "Let's talk about that."

My brow wrinkled as I asked, "What about it?"

Halix took a step back and circled the room slowly. His eyes scanned every inch as if he could see through the walls. "Bitcoin has become a . . . problem for GovCore. A problem that needs solving. And you two . . ." He paused again so the gravity of his words would sink in; "You two seem to have a particular interest in it."

I clenched my jaw, realizing that I needed to keep my composure. "We're not the only ones. Millions of people are into Bitcoin. Why single us out?"

Halix turned to me with an impassive scowl. "Because you don't belong with the rest of them. You're a part of the Collective, Toby. But you've gone off course. And GovCore doesn't like it when people go off course."

"Off course?" Ace echoed. His voice was shaky. "We haven't done anything. We're just living our lives."

Halix still not willing to look at him. "Your pathways are starting to look shaky. We have metrics on everyone, Author You haven't been to school in months; how does that qualify for the Ascendant Pathway. Toby, you haven't been to work. Your lifestyle is starting to raise eyebrows."

I stepped forward, now speaking in a firmer tone. "So what? Since when does taking a break make us criminals?"

Halix's smile was cold, mechanical. "Since you stopped following the rules."

The air in the room was thick with tension, both of us stood our ground. For a brief moment, there was a flicker of something in Halix's eyes. Was it recognition? Or regret? I wasn't sure, but it passed as quickly as it appeared.

One of the agents called out, "No signs of Bitcoin transactions, nothing in the computers."

Halix didn't turn around. "Good. Finish the search."

Ace and I exchanged a glance, a moment of relief passed between us. Maybe we were in the clear. Maybe we had managed to stay off their radar after all.

But then Halix turned back to face me. His expression was sinister. "One more thing, Toby." He stepped closer, and his voice dropped to nearly a whisper. "Did you buy a plane?"

My heart skipped a beat, but I didn't answer. Before I could open my mouth, Halix held up a hand. His cold gaze locked with mine. "Never mind. I'll see you soon."

Without another word, he turned and walked toward the door; his team of agents followed closely behind him. The apartment felt colder after they left, the tension lingered in the air. I stood there, frozen, watching as they disappeared down the hall.

Ace let out a long breath and sank onto the couch. "Man, I thought they were going to arrest us."

"Me too," I muttered, still staring at the door. "But something's not right. They didn't find anything, but . . ."

"They know," Ace finished for me. "They know something."

I nodded slowly, but my mind was racing. "They didn't come here for a routine inspection, Ace. They came here to send a message."

Concern was etched in Ace's face as he studied me. "So what do we do now?"

"We come up with a plan," I said, though my voice was less confident than I'd hoped. "And we get ready. This isn't over."

## THE CALM BEFORE THE STORM

As soon as the door closed behind Halix and his agents, Ace and I stood in silence. The aura of the inspection lingered in the air. The apartment suddenly felt foreign, like we were being watched even when we couldn't see anyone. My mind raced—there had to be bugs, hidden cameras, something. They hadn't found anything, but that didn't mean we were in the clear.

"Do you feel that?" Ace muttered, his eyes scanning the room like a hunted animal.

I nodded. "Yeah, I feel it. This place is probably tapped. We need to talk, but not here."

Without another word, we walked into the bathroom. I turned on the shower, so the sound of the rushing water could drown out any possible listening devices. Steam filled the room quickly, and the mist-covered mirror hid Ace's anxious face.

"Halix is onto us," Ace whispered, his voice barely audible under the noise of the water. "He didn't find anything this time, but you heard what he said."

I closed my eyes and rubbed my forehead as the thoughts of it all sat heavily on my brain. "We're screwed if we stay here. We need to get out of the country."

Ace's eyes flicked toward mine, wide with anxiety. "Get out of the country? Like . . . for good?"

"What choice do we have?" I shot back; my voice low but intense. "GovCore is tightening its grip, and they're not going to stop until they have us. Halix didn't come here for a routine inspection—he's got a file on us, Ace. He knows about the plane, the trips, everything."

"So, what's the plan?" Ace asked. He tried to keep his voice calm, but he failed. "Just up and leave?"

"Yeah," I said, pacing the small bathroom. "We've got enough money to live anywhere in the world. We need to disappear before they drag us into the Disruptive Crucible or even the Eclipse Circuit."

Ace winced at the mention of the Eclipse Circuit—the GovCore's solution for those who resisted them. Imprisonment was only the beginning. Once you were in the Eclipse Circuit, you were erased from society; your freedom was gone forever. I couldn't let that happen. We had to escape before it was too late.

"But until we can figure out a way to leave, we need to play it cool," I continued. "We have to act normal, pretend like we're still part of the system. You need to get back in school, and I'll try to get my job back."

Ace looked at me with concern in his eyes. "Toby, you really think they'll let us back in? After everything?"

"They have to," I said, although I wasn't sure I believed it. "We haven't done anything to get pulled out of our pathways yet. We're not on the criminal track—at least, not officially. We need to buy time until we can get out of here."

Ace sighed, nodding reluctantly. "Alright. We'll go back to work, school . . . act like nothing's wrong."

We stood in silence for a moment while the reality of our situation registered. We had gone from living the high life, flying under the radar, to now being hunted. I could feel the walls closing in.

I turned off the shower and opened the bathroom door. "Let's get some rest. We'll figure the rest out tomorrow."

Ace followed me out, and we moved into our separate bedrooms. I lay down on the bed, stared up at the ceiling, my mind racing. I tried to sleep, but the thought of Halix's cold stare kept gnawed at me. He knew. He might not have said it directly, but he knew something was up. The plane, the trips,—it was all adding up, and it wouldn't be long before GovCore made their move.

Suddenly, a noise broke through the quiet of the apartment. At first, I thought I was dreaming, but then it came again—heavy footsteps. I sat up in bed, my heart racing, and before I could even get up to reach for the door, I heard Ace yell from his room.

"Toby! You hear that?"

I jumped out of bed and moved quickly toward the sound of the door being forced open. The next thing I knew, the room was flooded with dark figures whose eyes glowed red piercing the darkness of the room. These AI-infused agents were more menacing than any I'd seen or heard of. Panic surged through me, but before I could react, they were on me.

A rough hand grabbed me from behind and forced a blindfold over my face. I struggled, trying to fight them off, but it was pointless. These weren't regular soldiers—these were the elite, trained to capture with precision. Within seconds, I was immobilized, and I could hear Ace engaged in his own struggle nearby.

"No! Let go!" Ace shouted; his voice muffled as they overpowered him.

We were dragged out of the apartment and shoved into a waiting vehicle. I felt the cold metal of handcuffs dig into my wrists. My heart pound against my chest. This wasn't how it was supposed to happen.

As the vehicle rumbled forward in the silence. I felt the warmth of multiple bodies near me; after some time, we stopped, and I was taken from the vehicle and placed somewhere else. We started moving again, and then I felt the unmistakable sensation of leaving the ground. We were in a plane. They had taken us in the middle of the night, no explanation, no warning. I could hear Ace breathing heavily beside me, but I couldn't speak .

Suddenly, the blindfold was ripped off. My vision adjusted, and there, seated in front of me, was Halix. His eyes were still as cold as ever, and there was a slight curve to his lips—a smirk.

"Didn't I say I'd see you soon, Toby?"

Ace was sitting next to me, bound and furious. His face was a mixture of fear and defiance.

"What the hell is this?" I spat, glaring at Halix. "Where are you taking us?"

Halix leaned back in his seat, and his mechanical eyes gleamed in the dim light of the plane. "Elara wants a word with you. Both of you."

"Elara?" Ace said; his voice was tight with anger. "What does she want with us?"

"You'll find out soon enough," Halix replied calmly, and turned his attention to the window as the plane soared through the night sky. "For now, just sit back and enjoy the ride."

And with that, the realization hit me like a punch to the gut. Elara Kestrel, the woman who had been pulling the strings all along, wanted us. And this time, there was no running away.

# CHAPTER 11:
# BEHIND THE CURTAIN

The flight had been silent; the steady hum of the engines kept me from unraveling. Ace and I exchanged glances, neither of us knowing what to expect when we landed. GovCore Headquarters was somewhere unknown, an undisclosed location off any maps, and yet here we were, being taken straight to the heart of it. For what? Neither of us could imagine.

When the plane touched down, the door opened, and the first thing that hit me was the crispness of the air. There was no chaos, no armed guards pointing weapons at us. Instead, a sleek black car awaited; its glossy surface reflected the artificial lights of the hangar. The agents ushered us inside without a word.

As we drove through the complex, the vastness of GovCore's headquarters became evident. Massive, imposing structures loomed in the distance, all sharp lines and gleaming glass. Everything about the place screamed power and control, and yet it was eerily quiet. Too quiet.

Finally, the car stopped in front of one of the taller buildings. We were escorted inside and led down pristine corridors, the floors so polished you could see your reflection. No one spoke a word to us, not even Halix, who had become a shadow since we left the plane.

And then, just like that, the hostility evaporated.

As soon as we stepped into what seemed like a private suite, the mood changed. The tension lifted as we were greeted by two attendants dressed in impeccable white uniforms. They led us to luxury quarters fit for royalty. The room was immaculate—rich dark woods, sleek marble, and floor-to-ceiling windows overlooked a perfectly landscaped garden. The scent of fresh flowers filled the space, contrasting sharply with the clinical feel of the headquarters outside.

"Is this . . . for us?" Ace whispered. He eyed the room as if it might dissolve into nothingness at any moment.

I shrugged, equally baffled. "Seems like it. For now, at least."

The attendants handed us fresh clothing, soft fabrics and tailored suits that felt impossibly expensive. They didn't say a word, they simply offered us a nod before leaving us to clean up after our long journey. I turned on the shower and let the hot water run down my back as I tried to process everything. Why were we being treated like this? If GovCore had really wanted us out of the picture, this wasn't how they'd do it. Something bigger was at play.

Once we were dressed, we were led to a dining room that was as extravagant as the quarters. A five-course meal was laid out in front of us—food that could rival any high-end restaurant. It was served on crystal plates so shiny you could mistake them for diamonds.

As we sat at the long dining table, I leaned toward Ace, keeping my voice low. "This doesn't feel right."

Ace nodded, while he picked at his food. "It's like they're buttering us up. Why the hell would they go through all this trouble if we're not in danger?"

I glanced around and noted the ever-present cameras that dotted the corners of the room. "Maybe they want something from us. If they were gonna arrest us, they'd have done it already. We would have been taken straight to the heart of their operation—GovCore doesn't do that for no reason."

Ace's eyes darted toward the door as he took a sip of his drink, his eyes darting toward the door. "So, what do you think they want?"

"Bitcoin," I said. The word came out before I could stop it. "They want Bitcoin, or at least control of it. They've seen how big this thing has gotten. The Bitpocalypse is spreading faster than they can contain it."

Ace frowned and leaned back in his chair. "You think they'll offer us a deal?"

I shrugged. "Maybe. But whatever they offer, it won't be good for us. They've already tried seizing Bitcoin from other people, but it's decentralized. They can't shut it down the way they want to."

We fell silent for a moment. Neither of us said it out loud, but we knew—GovCore was trying to figure out if we were a threat. And if we were, we'd be taken out, no questions asked.

"They want something bigger," Ace quickly whispered.

I nodded. "They always do."

After dinner, we were given more time to rest. The luxury felt like a distraction, something to keep us calm before the storm. We spent the evening in our quarters, speaking in hushed tones, knowing the walls had ears.

"Think about it," I said. I was staring out the window into the artificially lit garden. "If we weren't valuable to them in some way, we'd already be forgotten and relegated to the Eclipse Circuit."

Ace leaned back. His brow was furrowed again. "I keep thinking . . . what's their game? They're not going to let us waltz out of here. So what the hell do they really want?"

"I don't know," I replied. Tension rose in my chest. "But we've got to stay sharp. We're in the lion's den now."

The night wore on, and we tried to sleep, but it was restless. My mind kept spinning. What was Elara Kestrel's plan? Why bring us here and offer us all these luxuries, only to make a move later?

Just before dawn, a knock at the door interrupted the quiet. The attendants from before entered. Their faces unreadable. "It's time," one of them said; their voice was soft but firm.

Ace and I stood and exchanged one last glance of quiet understanding. Whatever happened next, we'd face it together.

The attendants guided us out of the dining quarters and down a long series of narrow corridors. The atmosphere shifted as we moved further into the belly of GovCore, where the environment felt more clinical than corporate. Every hallway was bathed in cold, artificial light, and I couldn't shake the feeling that we were being led into a labyrinth that was designed to disorient us. Even the air seemed heavier, more controlled, like everything within this facility was calibrated for maximum efficiency—and control.

Finally, after what felt like an eternity in twisting corridors, the double doors in front of us slid open, revealing a grand, dimly lit room. The ceilings were high, the walls sleek and metallic. And there she was—Elara Kestrel—standing at the far end of the room; her figure was backlit by soft, ambient light.

She was as striking as ever, maybe even more so. Dressed in a tailored suit that hugged her curves in all the right places, her hair fell effortlessly past her shoulders. Something about her commanded attention, not just because of her position of power, but because of the way she moved; the way her eyes seemed to cut through you. As we entered the room, her eyes landed on me, and I felt that familiar charge pass between us.

"Toby," she greeted me warmly with a soft smile. "And Author. I'm glad you're both here."

Before we could respond, she turned to the attendants and the agents who had escorted us. "You can leave now," she said. Her tone was polite but firm. "Everything is under control."

The room cleared out swiftly, and the doors slid shut behind them, leaving only the three of us in the massive space. Elara gestured toward the opposite end of the room where another door slid open, leading into another corridor. She motioned for us to follow her.

"This way," she said, in a casual tone, as if we were simply guests being given a tour.

As we stepped into the new hallway, the lighting shifted. It became warmer, almost inviting. Ace's eyes widened as we passed through rooms filled with sleek, futuristic technology. Advanced robotics, holographic displays, AI-driven systems—all things that would have been cutting edge even in the most sophisticated labs.

Ace's interest was piqued immediately. His gaze darted from one piece of equipment to the next. Curiosity pulled him deeper into this newly found world of innovation.

"You're welcome to explore," Elara said with a knowing smile, waving him off into one of the tech labs. "Take your time. Ask the engineers anything you'd like."

Ace didn't need another invitation. He was off in an instant. His eyes gleamed with excitement as he rushed into one of the labs. The doors slid shut behind him, leaving me and Elara standing alone in the corridor.

The second the door closed behind Ace, the mood between Elara and me shifted. The tension was palpable, a mix of anticipation and uncertainty. She stepped closer, reached out to me, and her fingers brushed against mine. I hesitated at first, unsure of her intentions, but then she took my hand firmly, her grip was soft yet possessive.

"It's been a long time, hasn't it?" she asked with a tone that was low and smooth, almost seductive.

I nodded, feeling the warmth of her hand against mine. "Yeah. It has."

Her eyes locked into mine. "I've always wondered about you, Toby. You were . . . different, even back then."

I wasn't sure how to respond to that, but the way her fingers tightened slightly around mine told me she wasn't making small talk. There was more to this conversation, more than just reminiscing.

As we began to walk, her hand remained locked with mine, and with every step, I could feel the connection between us deepening. It wasn't about physical attraction—though that was certainly there—but something more dangerous. I had seen enough to know that Elara Kestrel wasn't someone who simply wanted something. She needed it.

We turned a corner, and the door opened into another room—this one was larger, with transparent walls that gave us a panoramic view of the GovCore headquarters. Below us, thousands of people moved like clockwork, the perfect image of a finely tuned machine.

Ace, completely enraptured by the technology, had wandered off into another room, leaving Elara and me alone again. She stopped walking and turned to face me. Her hand still gripped mine, but this time it was tighter, more insistent.

"You've always been stubborn," she said with a hint of amusement. "I remember the first time I came to your house. You didn't trust me then, did you?"

I shook my head, my pulse quickened at the memory. "No, I didn't."

A small smile played at her lips. "And yet here we are, years later."

There was a beat of silence before I felt her thumb stroke the back of my hand, an intimate gesture that sent a shiver down my spine. "You've changed," she whispered. Her eyes darkened as they locked onto mine. "But in some ways, you're still the same."

I wasn't sure what to say. The heat between us was undeniable, and the way she held my hand, the subtle caress of her fingers, made it clear that this wasn't about business anymore.

"Maybe," I replied.

Elara's smile widened slightly. "Do you trust me now, Toby?"

"I don't know," I admitted, watching her closely. "Do I have a reason to?"

She stepped even closer, her hand still holding mine, but this time, I felt her other hand brush against my arm. "Trust is a funny thing, isn't it?" she murmured. "It takes years to build, but only a second to lose."

Before I could respond, she gently pulled me forward, led me across a long, glass bridge that stretched over what appeared to be the central hub of the GovCore headquarters. The bridge offered a dizzying view of the entire complex. I could see everything—miles of technology, workers, agents, and engineers, all moving in sync like parts of a massive organism.

We approached the enormous hive-like structure at the far end of the bridge, and as we neared it, Elara finally released my hand. The loss of contact was immediate, but the tension between us remained thick.

"This," she said, gesturing to the hive structure, "is the heart of GovCore. It's where everything happens."

I couldn't take my eyes off her. There was something magnetic about her presence, something that made it hard to pull away. But I couldn't forget who she was—what she represented. Elara Kestrel wasn't just any woman. She was dangerous.

"Come with me, Toby," she said, motioning toward the entrance to the massive structure. "This is where the real power resides."

I hesitated for a moment. The familiar tension knotted in my chest. The pull of her offer was intoxicating, but I knew better than to trust anyone from

GovCore—especially Elara. Still, something about her allure, the dangerous promise in her eyes, made it impossible for me to turn away.

"This place," she continued, in a whisper as if she were sharing a secret, "we call it The Nexus."

The name made the hair on my arms stand up. It sounded ominous—appropriate for something that held the kind of power GovCore wielded. With a single glance at Ace, who was still trailing behind, wide-eyed with curiosity, I followed her into The Nexus.

The moment we stepped inside was like walking into the heart of a futuristic dream—or nightmare. The air was filled with the low-frequency sound of hundreds of machines processing terabytes of data at unimaginable speeds. The walls were lined with rows of screens that displayed cryptic streams of code, complex patterns, and swirling lights that seemed to pulse in time with the rhythm of the artificial intelligence controlling it all.

At the center of the room stood the true brain of GovCore: Neutrinon, the ASI—the artificial super intelligence that ran every operation within the global network. Its presence was suffocating, not because of physical intimidation but because it knew *everything*. I could feel it watching us, processing every move, every twitch of my muscles, every micro-expression on my face.

"Neutrinon sees everything," Elara said, glancing at me with that dangerous glint still in her eyes. "It's more than artificial intelligence. It's the mind behind the entire system."

Ace trailed in stunned as he absorbed the sight. He was lost in the technological marvel, mesmerized by the sheer power of the machine. I saw it in his eyes—the awe, the thirst for knowledge that had always driven him.

Elara approached him and gently guided him toward the massive console. "Have a seat, Author. Neutrinon would like to ask you a few questions."

Ace, still in a trance, obliged without hesitation. He sat down in the sleek chair that seemed to mold itself to his frame, facing a holographic display of the ASI's interface. The moment he was seated, the screens around us shifted. They now displayed layers of digital architecture and streams of encrypted data. The ASI's voice crackled through the room, smooth and emotionless but with an eerie presence that felt like the intro to a horror movie.

Author," it intoned, "you are an exceptional mind. You have studied deeply the intricacies of artificial intelligence, blockchain, and cryptography. You know much."

Ace nodded cautiously, glancing back at me for reassurance. I gave him a slight nod. "I know a few things," he replied, trying to keep his tone neutral.

Neutrinon paused for a second, as if digesting Ace's answer. Then the holographic display shifted and showed a stream of information that I immediately recognized—blockchains, Bitcoin transactions, and data flows from decentralized exchanges.

"What do you know about Satoshi Nakamoto?" the ASI asked. Its voice took on a sharper edge.

Ace stiffened, and I felt a jolt of panic surge through me. Neutrinon had already begun connecting the dots. It had combed through the deepest corners of the dark web, piecing together the origins of Bitcoin. The ASI was closing in.

Ace swallowed hard but kept his composure. "Satoshi Nakamoto is a pseudonym," he said carefully. "No one really knows who he is. It's . . . a mystery. Always has been."

Neutrinon's voice grew cold, almost metallic. "But you were Satoshi's dorm mate. You had access to early discussions about Bitcoin. The original white paper—where is it located?"

The room seemed to close in around us. Ace's face paled, but he kept his voice steady. "Yeah, Satoshi and I were close at one point. But we lost touch. I don't know where he is now."

There was a beat of silence. I could feel Neutrinon probing deeper, analyzing Ace's body language, searching for inconsistencies, reading every nuance of his voice for deception.

Elara stepped closer to me, and her hand brushed mine again. Her fingers wrapped around my palm. The electric touch felt both comforting and insidious. Her eyes glanced between me and Ace, and then she whispered, "This could all be yours, Toby. Ours—together."

Her words felt like a seduction; a promise bathed in deception. *Together?* I thought. My mind reeled. Was she offering me a place in this madness, to be by her side as GovCore's reign continued?

I didn't have time to answer because Neutrinon's voice interrupted, now with a more hostile undertone. "Author, it is in your best interest to tell the truth. Satoshi Nakamoto's disappearance correlates too closely with your movements. Where is he?"

Ace, being the quick thinker he was, leaned into the console; his face was calm, but I could tell he was on edge. "I'm telling you the truth. I don't know where he is. All I know is what's already out there."

Neutrinon paused again. The air was stifling, the unease was palpable. I could feel Elara's eyes on me, her thumb gently caressing the back of my hand as though reminding me of the promise she had made. But all I could think about was getting out of there. Neutrinon was relentless, and Ace couldn't keep dodging its questions forever.

As I stood next to Elara, the subtle seduction in her words seemed to slip into something darker. The line between ally and enemy had never been more blurred.

"Tell me, Toby," she whispered, leaning in closer. "What do you want? Power? Freedom? You could have everything at your fingertips. You and Author."

I closed my eyes. I was torn between the warmth of her hand in mine and the cold reality of what GovCore truly represented. This was just the beginning.

The room was silent. When Ace stood up from the chair his face was pale but composed. Neutrinon had pressed hard, but Ace had held his ground, dodging its questions like a seasoned chess player evading checkmate. I watched him as he walked past me, offering a slight nod, but the stiffness in his shoulders was evident.

My turn.

"Mr. Toby Valorian," Neutrinon's voice echoed through the room, now focused entirely on me. Its tone shifted—cold, precise, and unnervingly neutral.

I was guided to the chair with my heart pounding as I was seated. The sleek, metallic surface of the console felt cold beneath my fingertips. The room seemed to narrow, and I became acutely aware of Elara Kestrel's presence behind me. Her hands gently rested on the back of my chair, and her fingers lightly brushed against the fabric, as if she could feel the anxiety building in me.

The ASI's face on the holographic display shifted; its features smoothed out into something more neutral, less hostile, but still calculating. It was like I was being sized up and broken down into data points and code. I took a deep breath, trying to steady myself, and then the first question came, simple but weighted.

Neutrinon's digital eyes were locked onto mine when he asked, "Do you know who your father is, Toby Valorian?"

*I was caught off guard. My father?* The question lingering like a ghost I hadn't thought about in years. "No," I said. "I never knew him."

Neutrinon paused, as if calculating my response, before continuing the questioning.

"It appears you have accomplished much despite not knowing your true lineage. Let's review," the ASI continued; its tone was indifferent but precise. Suddenly, my life unfolded on the display as if my entire existence had been reduced to a list of achievements and failures. The gun situation during the inspection when I was a child. My military experience, narrowly surviving combat. Refusing GovCore's special offer. Landing a high-paying consulting job despite the odds stacked against me. And then Bitcoin.

Bitcoin.

I tensed in my seat as Neutrinon shifted its focus. It began to ask questions, similar to the ones it had thrown at Ace. My connection to Satoshi. How Bitcoin had risen from an obscure digital experiment to a financial revolution that threatened GovCore's control. But I had rehearsed these answers in my head a thousand times.

I stuck to the truth, or at least parts of it. *I deflected, sidestepped.* I revealed just enough to keep Neutrinon from digging deeper, but not enough to implicate myself. I told it what I *knew*, but not what I *felt*.

As the questions rolled on, there was a subtle shift in the air. I could feel Elara's gaze burning into the back of my head, her fingers tightening slightly on the chair as the ASI continued its barrage. And then, out of nowhere, Neutrinon threw a curveball.

"You may be more connected to GovCore than you realize," it stated, its digital eyes narrowed slightly as if trying to peer into my very soul. "Your bloodline . . . there may be something . . . within you."

I froze. My heart skipped a beat. My bloodline? What the hell was it talking about? It didn't elaborate, just left the implication hanging in the air like a snake poised to strike. But there was no confirmation, just a hint, a suggestion that there was more to my past than I knew.

The room felt colder after that, and Neutrinon seemed to pull back. Its questioning finally came to an end. "You are dismissed, Valorian."

I let out a slow breath, realizing how tense I'd been throughout the interrogation. The ASI had pried, but it hadn't broken me. I stood up with rubbery legs and turned to find Elara watching me closely. Her expression was unreadable, but there was something in her eyes—something almost . . . proud?

"Follow me," she said softly, motioning toward the exit. "There's more we need to discuss."

Ace gave me a confused look. He was rattled with anxiety as he turned to head back to our quarters. I watched him go, then I turned to follow Elara. My mind raced. Had Neutrinon revealed the truth? Or not.

As we walked through the winding halls of GovCore's headquarters, I couldn't shake the feeling that the more I uncovered, the deeper I was being pulled into something I was unsure about but wanted to understand. Elara led me through the maze with silent grace. her hand grazed my arm every now and then, a subtle reminder that she was still in control.

And now, I was being pulled into her orbit. Whether I liked it or not.

# CHAPTER 12:
## BLOODLINE OF POWER

The car ride was silent, except for the sound of the engine gliding along. Elara sat beside me with legs crossed elegantly, eyes forward, as if she had already mapped out every second of what was to come. I was still reeling from everything that had happened—Neutrinon's interrogation, the strange offer, the flirtation, the tension. My mind flooded with thoughts of what the future might hold, but I didn't expect what was waiting for me next.

We pulled up to an immaculate estate within the GovCore premises. Huge, majestic-like, with high security everywhere. The type of place where you imagine only the untouchable live. And yet, here I was, stepping out of the car and walking into the belly of the beast.

"This is where I live," Elara said in a voice that was enticing but firm. Her words lingered as I looked around at the pristine landscapes, the shimmering glass windows, and the presence of armed guards stationed discreetly, yet visible enough to remind you of who was in control.

"What do you mean by . . . this could all be mine?" I asked, with curiosity and caution in equal measure.

She smiled a slow, deliberate smile that conveyed more than amusement. "I'm going to show you," she said, leading me inside.

The conversation shifted, flowing easily between us as we spoke about GovCore, power, the future. But I could feel something darker beneath the surface. Her words, the way she looked at me—everything was carefully calculated, like she had already anticipated every move. Then she slipped away to freshen up, and when she returned, she was wearing something far more provocative. Her intentions were clear.

We ended up in her bed. The night was a whirlwind of passion and seduction—intense, raw, erotic. The kind of night that leaves you questioning everything the morning after.

When dawn broke, the world felt different. I woke up next to her, and the reality of what had happened sank in. Elara was sitting across from me, calm, composed, like she hadn't torn apart my world with one night. She looked at me. Her eyes narrowed slightly, and then she dropped the real bombshell.

"Do you want to know who your father is?" she asked in a silky smooth voice. And yet there was a sharp edge to the question.

I froze, unsure of what to say. She slid an envelope across the table toward me. My hands felt heavy as I opened it, there were pictures inside. I stared at them struggling to make sense of what I was seeing.

Pictures of a man and my mother. Together. The man looked familiar in a way I couldn't explain.

"His name is Paul Kestrel," she said; her voice was barely above a whisper, but it cut through the air like a sharp sword. "He's your father, Toby."

My stomach twisted; my vision blurred as I stared at the photos. Paul Kestrel. Elara's father. My father. I felt sick. I had just spent the night with my half-sister, and she knew it all along.

I looked up at her feeling nausea. She watched me with a smile, as if she took pleasure in the torment that was tearing me apart from the inside.

"You knew," I said.

"Of course I knew," she replied. That twisted smile never left her face. "The bloodline has to stay pure. It's the only way GovCore can survive into the next era."

I could barely comprehend the words. Bloodline. Purity. GovCore. It was all connected, all part of a plan I hadn't seen coming. And now, I was in the middle of it.

Suddenly, I noticed the heavy presence of security approaching the building. My pulse quickened. I knew I was standing on a precipice—one wrong move, one wrong word, and this could all go very, very wrong. I tried to figure out what to do next.

"I need to think," I said quickly, trying to buy time. "I need time to process this. What about Ace?"

Elara's expression softened slightly although the calculating gleam in her eyes remained. "He's on the ascended pathway. He's your brother. He'll be well taken care of. He'll go on to do great things. We'll all be part of something greater."

I forced myself to stay calm, to keep my voice even as I said, "Give me some time. Let me think."

She studied me for a moment before nodding slowly. "Fine. But don't take too long, Toby."

Her words were a warning. A subtle threat.

As I was driven back to pick up Ace, I tried to make sense of what I had learned. Elara left me with a final message as we boarded the plane back to Atlanta. "I'm waiting," she had said; her voice echoed in my mind long after we took off.

Ace sat beside me, oblivious to the storm raging inside my head. The plane soared into the sky, I watched GovCore as we left it behind. But I knew—this was far from over. The game had truly begun.

# CHAPTER 13:
## SEEDS OF REBELLION

As we touched down in Atlanta, the familiar aura of the city wrapped itself around me like a weighted blanket. The skyline was the same, but everything felt different now. GovCore had infected my mind like a virus. Its tendrils clinging to every thought, every move I made. I hadn't spoken a word about what had happened at their headquarters. Not to Ace. Not to anyone.

Ace noticed the change in me immediately. We exited the corporate jet, and even though we tried to slip back into the normalcy of our apartment, it was clear I wasn't the same. Elara's revelation—of my true identity—sat heavy on my heart. I couldn't shake it. I couldn't bring myself to tell Ace that we were more connected to GovCore than we ever thought possible. I didn't want to burden him with that.

"You've been different since we got back," Ace said one evening as we sat in the living room. The glow of the TV bathed the room in dull light, but I wasn't paying much attention. I'd been staring at the same screen for hours. My mind was miles away.

"I'm fine," I muttered.

"Bullshit," Ace said. He stood and moved closer to me. "You haven't been fine since we left GovCore. What happened back there? What did Elara say to you?"

I shook my head, unwilling to dive into the depths of what I'd learned. "It's nothing we need to talk about right now. The important thing is what we're going to do next."

Ace didn't push. His eyes told me he wasn't satisfied. He sat down, but his curiosity lingered between us like smoke.

"So what now?" he asked after a long pause. "We've got all this Bitcoin, but if we stay quiet and sit on it, it won't do anyone any good. People are struggling out there, and the Bitpocalypse is only getting worse."

He was right. Bitcoin had given people a taste of freedom, and GovCore had retaliated with an iron fist. But there was still a way to fight back without falling into the traps GovCore had set. We had to be smarter than them.

"We need to form a group," I said, my voice low but resolute. "A group of pro-Bitcoin people who want to do it the right way. By the rules. By GovCore's own standards. We help people use Bitcoin legally, without hiding, without falling into GovCore's clutches."

Ace's eyes lit up, and he leaned forward. "Like a movement?"

I nodded. "Exactly. We show people that Bitcoin isn't for gamblers or for those who want to break the law but a tool that can be used by everyday people to better their lives. We can help them pay taxes, convert their earnings, do everything the right way. And we stay under the radar."

Ace smiled. "That could work. But how do we find these people? We can't exactly advertise it on the news."

"We don't need to," I replied. "People are out there, looking for a way to stay on the right side of the law while using Bitcoin. We have to reach out in the right places. Forums, underground communities, even people we know personally. Like that guy we met—what's his name—Don, the Navy pilot."

Ace's mood began to lighten up. "Don . . . yeah, he was asking about how to do things legally. He has some Bitcoin stashed away and he doesn't know what to do with it. Plus, he wants to take care of his parents."

"Exactly," I said, leaning forward. "He's our first recruit. Then we help him, and once we do that, word will spread. People will come to us."

We spent the next few days reaching out quietly, meeting with Don and helping him navigate the murky waters of Bitcoin regulations. He was a good guy—ex-military, with an air of discipline and duty about him. He didn't want to hide his Bitcoin or use it for anything shady. He just wanted to convert some of it to cash, pay his taxes, and make sure his parents were taken care of.

"I don't need millions," Don said during one of our meetings. "I just want to use what I've got to help my folks. They're getting old, and I don't know how much time they have left. But GovCore's breathing down everyone's necks. I don't want to end up on the wrong pathway."

We guided him through the process, slowly building up trust. Over the next few months, our group started to grow. It wasn't a massive movement—just a handful of regular, working-class people who saw Bitcoin as a way out of the Collective grind.

Ace and I remained in the shadows. We never revealed how much Bitcoin we had, or the fact that we had been involved from the very beginning. But we championed Bitcoin from the sidelines, helping others use it the right way. And even though we stayed under the radar, it became clear that GovCore wasn't going to ignore the growing number of people using Bitcoin legally. The tension was rising, and we knew we were running out of time before things took a darker turn.

"How long do you think we can keep this up?" Ace asked one day after another long meeting with our group.

I stared out the window, watching the lights of the city twinkle below. "As long as we have to," I said. "Until we figure out the next move."

But even as I said it, I could feel the force of GovCore's gaze bearing down on us, waiting for the perfect moment to strike.

We weren't hiding anymore. We were on the front lines. And it was only a matter of time before GovCore came knocking again.

## THE DANCE WITH ELARA

The days passed, and while Ace and I were busy building our group of Bitcoin advocates, life outside those meetings seemed to take on a strange rhythm. Every few days, like clockwork, I'd find a new package waiting at our door. Some were rather small trinkets—an expensive pen, a fine bottle of whiskey— but others were more extravagant: a custom-tailored suit, a watch that probably cost more than most people made in a year. And every time, it came with a note.

"Have you made up your mind up?"

-Elara.

It was almost playful, the way she kept reaching out. As if our last encounter at GovCore had never happened. As if she hadn't revealed the dark, twisted truth of who I really was—her half-brother. As if she weren't watching my every move, waiting for me to come back to her. But there was no denying it— she was still there, hovering over my life like a shadow.

The text messages were just as frequent. Every night, my phone would buzz, and there she was, like a ghost in my pocket. We exchanged short messages, nothing too serious—at least not on the surface. She would ask how I was doing, how Ace was handling things, and I would respond with vague, noncommittal answers. But there was always an undertone in her words, a sense that she was still pulling strings, working angles, even if I didn't know exactly what they were.

"How's your group doing?" she asked one night in a text. "The Bitcoin one. I hear it's growing."

I didn't hesitate to respond. After all, there was nothing secret about the group. We were operating out in the open, making sure to follow every rule GovCore had laid out. We weren't doing anything illegal. At least not yet.

"It's doing fine," I texted back. "People just want to use Bitcoin the right way. Pay their taxes, convert it legally. Nothing shady."

I could almost imagine her smile on the other side of the phone. She loved to play these little games, always probing, always trying to figure out what my next move would be.

"Good," she replied. "I like it when people play by the rules. Makes things more . . . predictable."

There was a pause before her next message came through.

"And you? Still thinking about what I said?"

I didn't respond right away. How could I? She had laid everything out on the table, offered me a seat at the highest level of GovCore; told me that we could rule together, keep the bloodline pure. But the thought of it still made my stomach turn. I wasn't like her. I wasn't like them. At least, that's what I kept telling myself.

"Still thinking," I finally typed, and sent the message before I could second-guess myself.

Her response was almost immediate.

"Take your time. But don't take too long."

Even through a screen, I could feel the threat behind her words. Elara Kestrel was always one step ahead, always planning, always watching. I knew she had her eyes on me and Ace, and that she was waiting for the right moment to strike.

As I put my phone down, I couldn't help but wonder what she was up to. There was no way she was just sitting back, waiting for me to come to her on my own terms. No, Elara was cooking up something big, something dangerous. And I had to be ready for whatever came next.

But for now, I had my own plans. We were building something with our Bitcoin group—something real, something that could help people escape GovCore's stranglehold. And no matter how many gifts she sent, no matter how many times she reminded me of our connection, I wasn't going to let her pull me back in. Not yet.

# CHAPTER 13 (CONTINUED): THE INDEPENDENT PATHWAY

Don, always the pragmatic thinker, came to us one evening with a solution.

"I know a guy," he said, lighting up a cigarette. "He's sharp, real sharp. A lawyer, used to have a gambling problem, but he's clean now. I helped him out once, back when he was down on his luck, and now he wants to return the favor."

Ace and I exchanged glances. We trusted Don—he had been with us for some time now—but a lawyer with a shady past wasn't exactly what we envisioned for the group. Still, we needed the help, and if this guy could show us legal loopholes and strategies to help the group convert their Bitcoin into something more sustainable, we were all ears.

The next day, Don introduced us to Patrick "Pat" Malone, a tall, well-spoken man with a sharp suit and an even sharper gaze. He shook hands with the kind of confidence that only comes from a man who's clawed his way back from rock bottom.

"Don tells me you're looking to help people move out of the system," Pat said, his voice was calm but it had an edge. "To break free from GovCore's pathways, I can help with that. But you've got to be smart about it."

He laid It out for us, breaking down a series of intricate steps that would allow people in our Bitcoin group to not only convert their Bitcoin into real money but also to protect it. He spoke of trusts, offshore holdings, and business structures that could shield wealth from prying eyes. He talked about tax strategies, and how to navigate the legal system without drawing unnecessary attention.

"GovCore may have their pathways," Pat explained, "but the legal world is full of gray areas. If you know where to look, you can move between the cracks."

It was genius, really. Pat had found a way to build an independent pathway, one that didn't depend on GovCore's rigid structure. For the first time, our group had a real chance at escaping the system entirely. No longer part of the Collective, not quite in the Ascendant Pathway, but something completely new—outside of GovCore's reach.

As Pat continued to advise us, the group grew more confident. People who had been nervous about their newfound wealth—afraid that they would be taxed or jailed for trying to convert their Bitcoin—now felt empowered. They were setting up businesses, investing wisely, and ensuring that they had the means to live comfortably without relying on GovCore for anything.

"This is it," Ace said to me one night as we watched the group grow. "People are starting to see that they don't need GovCore. They don't need their pathways. They can be free."

But with every new step of independence, I knew we were getting closer to a dangerous line. GovCore wouldn't tolerate this for long, and Elara . . . I could already feel her presence, hovering over us like a storm cloud. Every gift she sent, every text message she sent asking if I had made my decision, was a reminder that she was watching.

I knew it wouldn't be long before GovCore noticed what we were doing. And once they did, they'd come down hard on us—especially Elara. She had made it clear that she wanted me to be a part of GovCore's future, not running some rebellious group that threatened their control.

"You know this is going to piss them off, right?" I said to Ace one day, as we discussed the group's progress.

He nodded with a determined look on his face. "Yeah, but it's the right thing to do. People deserve the chance to live free."

He was right, of course. But that didn't change the fact that we were walking a fine line. And as much as I wanted to stay focused on helping people, I couldn't shake the feeling that this was all going to blow up in our faces.

Pat's plan was brilliant, but it was also dangerous. The more people we helped break free from GovCore, the more attention we'd draw. And sooner or later, Elara was going to stop asking me if I had made my decision—and start demanding it.

While the group continued to thrive, I knew the clock was ticking. We had found a way to help people escape the system, but GovCore wasn't going to let us do it without a fight. And when that fight came, it wasn't going to be only me and Ace on the line—it was going to be everyone we had helped.

This new Independent Pathway might have been the key to freedom, but I knew it was also the key to our undoing if we weren't careful.

## DECISION

The day had finally come—Elara Kestrel gave me the ultimatum. It was now or never. Either I was in—standing beside her at GovCore—or I was out, set on a more grim pathway. I had no plan—nothing concrete to fight back with. The independent pathways we'd created with Don, the Navy pilot, and Pat the attorney had given hope to people trying to escape GovCore's grasp, but it wasn't enough to topple the behemoth.

And I knew it.

Ace and I sat across from each other in the small apartment that had, for a while, been our sanctuary. The city outside was the only sound between us as I struggled to find the right words. He was already looking at me with suspicion, skepticism, like he knew what was coming before I said it.

I broke the silence. "I'm going back in," I said.b

Ace shot up from his seat; his eyes bulging in disbelief. "What do you mean you're going back in? You can't trust her, Toby. She's playing you. You know that!"

I shook my head, trying to keep my voice calm. "I don't have any other choice right now. We don't have the resources, the numbers—nothing. The only way we can make any of this work is from the inside. I have to get inside GovCore, into their systems, into their plans."

"And what? Be her puppet?" Ace spat, pacing the room. "She's going to chew you up and spit you out. You think you can outplay her? She's got Halix. She's got Neutrinon. You're walking into a trap."

"I know that!" I snapped. "You think I don't know that? But right now, she thinks I'm with her. She thinks I'm going to fall in line. And that gives me leverage. If I can gain her trust—if I can get deeper into GovCore, then maybe, maybe, I can find a way to end this from the inside."

Ace stared at me with a frown and balled fists. "So what? You're going to leave me out here to fend for myself? We're in this together, Toby."

I lowered my tone, trying to get through to him. "I need you here, Ace. I need you to keep everything stable. Keep the group moving forward; keep the pathways alive. If we all go in, we're putting everything at risk. We need someone on the outside who's still free, someone who can act if things go south for me."

He looked away. The hurt he felt was plain on his face. I couldn't blame him. I was asking him to sit back while I walked right into the lion's den. But this was the only way.

"I'm not switching sides," I added, trying to reassure him. "I'll keep you filled in. Every step of the way. Trust me, Ace. I'm not going to let her win. I'm going to handle this."

Ace let out a frustrated breath before finally turning back to me. "You better know what you're doing, man, cause if you get caught up in that web, it's not only you they'll take down. It's everything we've built."

I walked over to him and clapped a hand on his shoulder. "I got this. I promise you."

For a moment, we stood there in silence. There were no guarantees. No promises that this would work. But we had to take that chance. I had to take that chance.

And with that, I packed my things, ready to step back into the world of GovCore, ready to walk that tightrope between friend and foe, knowing that the future of Bitcoin, Ace, and everything we had fought for rested on my shoulders.

I had to play the game now. But I was playing it on my terms.

# CHAPTER 14:
# RETURN TO THE LION'S DEN

I stared at the text message from Elara Kestrel for what felt like an eternity. I had made my decision to step back into the world of GovCore, but it didn't make it any easier. When I sent her the message letting her know I was ready to join her, the reply was immediate—"I'll send a plane."

This wasn't a kidnapping like the last time. This time I was being welcomed back as one of their own, though I wasn't sure what kind of reception I'd get. Would she truly trust me? Did she buy into my charade? Or was this all part of her game?

Within hours, a sleek, black, private jet appeared at the small airstrip not far from where I was staying. As I approached, I expected the plane to be empty; I assumed that I'd have time to gather my thoughts. But as I stepped inside, I immediately saw her—Elara Kestrel, seated in a plush leather chair, a glass of wine in her hand, and an almost delighted expression on her face.

"Toby," she purred as I took a seat across from her. "I'm glad to see you've come to your senses."

I nodded and kept my expression neutral. "I figured it was time."

Her eyes sparkled with something—a mixture of amusement and something more insidious. "It's not often someone gets a second chance with me, Toby. You should consider yourself lucky."

I forced a smile, careful not to reveal the swirling chaos of thoughts inside my head. "I do."

The plane lifted off, and the sound of the engines filled the cabin. Elara leaned back, uncrossing her legs and studying me. "We have a lot to talk about," she said softly, "When we return to GovCore, I'll show you everything. Our future, Toby. Our plans. There are things you need to see—things only I can show you."

Her words were laced with a strange kind of promise, one that made me uneasy. But I stayed calm, nodding as if I were eager to know what came next.

"I'm looking forward to it," I replied.

She smirked. Her eyes gazing provocatively. "You should be. The world is going to change, and you and I, we'll be at the center of it."

As the conversation shifted, Elara slid out of her seat and moved to sit beside me. I felt her hand graze my arm; her fingers lingered on my skin as she moved closer. I tensed but kept my cool. As she pressed herself against me, her lips brushing against my ear.

"We could start now, you know," she whispered; her breath was warm against my skin. "We don't have to wait to get back to GovCore."

I forced a laugh; my voice smooth. "Elara, I think we both know there's more important business ahead of us right now."

She leaned back, studying me with a pout, but she didn't push it. "You're right," she said with a sigh. "We have all the time in the world for that."

Her lips found their way to my cheek, soft and deliberate, but I made sure to keep it at that. I couldn't afford to let this go too far. I needed to keep her on my side, but I also had to maintain control.

For the rest of the flight, the air between us was tense, a dangerous game of cat and mouse. Elara clearly wanted more, but I kept her at arm's length, deflected her advances with smooth words and carefully calculated gestures.

As the plane began its descent, she leaned back in her seat with a sly smile on her lips. "When we land, Toby, you'll see everything. And once you do, you'll understand why I need you by my side. Why we need each other."

I nodded imagining what awaited me back at GovCore. Whatever her plans were, I had to stay sharp, stay focused. The stakes were higher than ever, and the only way to win this game was to play it better than she did.

As the plane touched down, Elara reached out and took my hand; her grip was firm and possessive. I let her hold my hand, knowing that for now, I was playing a part. But inside, I was already planning my next move.

This was no longer a game. This was the beginning of a rebellion.

# CHAPTER 15:
# THE KING OF GOVCORE

As the wheels of the jet touched down on GovCore's private airstrip, the shift in the air was palpable. Elara gave me a lingering look; her smile was a mix of satisfaction and triumph. But while she believed she had me in her grasp, I was already planning. The plan was in motion, and I had to play this part perfectly. Every detail had to be precise.

The moment we stepped out of the plane, an army of security personnel, all dressed in black, stood in perfect formation, waiting for us. At the center of it all was Halix. His face was as stoic as ever. Elara waved her hand casually, signaling to them, "He's with us now. Toby Valorian is on our side." Her voice echoed with authority, leaving no room for doubt.

I could feel their eyes on me, but no one questioned it. It was preordained. They were already prepared for this, like actors waiting for their cue. I was being welcomed, not as a guest, but as something more—almost like a king stepping into his kingdom, side by side with the queen.

"Come," Elara said as she slipped her arm through mine. "There's so much to show you."

We moved quickly from the airstrip to an opulent convoy of cars that whisked us deeper into the GovCore fortress. As we traveled through the labyrinthine roads, Elara began gifting me lavish things—suits from the finest tailors, diamond watches, access to limitless funds, and my own security detail. It was all part of the show, part of the illusion. She wanted me laced in luxury, the pinnacle of GovCore's elite.

But while she thought she was impressing me, I was unmoved. I was already wealthy in my own right, thanks to Bitcoin. The gifts she was giving me were nothing compared to what I had amassed. Yet, I had to wear the mask. I had to play the role to perfection.

"We'll be living together," she said casually, as if it were another natural step. I nodded, not surprised, knowing this was part of the arrangement. I couldn't make her doubt me for even a second.

After getting cleaned up, we headed to the heart of GovCore—the boardroom, where the power of the world's most insidious corporation congregated. The walls were adorned with GovCore's insignia, and the table stretched across the room, surrounded by decision-makers, upper-echelon lieutenants, and, of course, the centerpiece of it all—Neutrinon.

Neutrinon's presence was as imposing as ever; its digital face flickered with the cold calculation of advanced artificial intelligence. It was the brain behind GovCore, the entity that knew and controlled everything.

All eyes were on us as Elara led me into the room and introduced me as her 'right hand, someone who would help her make decisions. "His military background, his bloodline—they speak for themselves. Toby will be an invaluable asset to us," she announced, and everyone at the table nodded in agreement, as if I'd been a part of this empire for years.

Then she turned to me. A sly smile crept onto her lips while she revealed the centerpiece of their plan. "Tomorrow, we're announcing the CoreCoin," she said, pausing to let the effect of her words sink in. "A centralized digital currency connected to the Cerebrax Seal. Elara's voice echoed through the room, confident and commanding, as she said, "And it will all be controlled by Neutrinon." The words lingered, but I wasn't letting them settle. Not yet. I needed more information—much more.

I leaned forward with my fingers interlaced, playing the role of the curious new strategist. "So, explain how it all works. CoreCoin is the replacement, but what exactly is the mechanism that ensures compliance?"

Elara glanced toward Neutrinon's digital form on the screen. It flickered for a moment before responding. The artificial intelligence's voice was calculated—almost human, but not quite. "CoreCoin is the next evolution of digital currency. It is fully integrated into the Cerebrax Seal, which every citizen will be required to have in order to convert their existing cryptocurrency holdings into CoreCoin."

My face remained neutral, unlike my swirling thoughts. They weren't just trying to replace Bitcoin. They were creating a system that would surveil, control, and restrict the flow of money down to the individual.

"And how exactly does the Cerebrax Seal work?" I asked, glancing at Elara for a moment before shifting my attention back to the AI. "What's the full process?"

Elara smiled, clearly pleased with my interest. "The Cerebrax Seal is an implant—it's both a physical and digital signature embedded within each individual. It links directly to Neutrinon's database, tracking not only every transaction but also the identity and activities of each user. The idea is to create total transparency—ensuring that no illicit or untaxed activity occurs."

Neutrinon's voice interrupted again, detailing the specifics. "Once an individual opts to convert their Bitcoin or any other cryptocurrency, they must submit to a Cerebrax scan. This scan verifies their identity through comprehensive biometric analysis, linking them directly to their digital wallet. Only those with the Seal can perform transactions."

I furrowed my brow. "And what if they don't want the Seal? What if they refuse to convert?"

Elara's expression remained calm, but there was a hint of something darker behind her eyes. "They won't have a choice. Eventually, all other forms of cryptocurrency will become obsolete, either due to regulation or outright bans. It will begin with incentives, but soon, CoreCoin will be the only recognized digital currency, and without the Seal, citizens won't be able to participate in the economy—legally."

One of the lieutenants, a man sitting across the table, chimed in. "The Seal will track spending patterns, alert us to unusual activity, and allow Neutrinon to flag individuals who may pose a risk to the system."

"A risk to the system?" I pressed.

When Neutrinon responded, its voice was chilly, detached. "Those who attempt to circumvent the CoreCoin system, engage in illicit activities, or who refuse compliance with the Seal may be classified as Disruptive. The pathways will adjust accordingly."

I nodded slowly, trying to process the gravity of it all. This was more than a currency—it was a tool for control. Every financial decision, every purchase, every transaction would be under the watchful eye of GovCore, and Neutrinon would be the omnipresent force behind it.

"And what about Bitcoin?" I asked, shifting gears slightly. "Is the goal to eliminate it?"

Elara leaned back and folded her arms. "Not immediately. For now, we let people believe they have a choice. CoreCoin will be introduced as a parallel currency, a safer, more convenient option. Bitcoin and other cryptos will still exist, but gradually, we'll introduce measures to tax, regulate, and ultimately phase them out. CoreCoin will be the only option left."

I tapped my fingers against the table. "So, we dangle the carrot first—the promise of security, ease, and wealth. But the stick comes later."

Neutrinon sounded pleased. "Correct," it responded.

The room fell silent for a moment, and my mind buzzed with the implications. True, this was about replacing Bitcoin. This was also creating a new era of surveillance—of absolute control.

I glanced at the others around the room. They were impressed, nodding as if my questions had exposed my value. But I knew the truth—I had asked those questions not because I wanted to help, but because I needed to know how deep this rabbit hole went.

Elara's gaze was locked onto me. Her smile widened. "I knew you'd understand, Toby. This is the future. This is what we'll build—together."

I turned and forced a smile. "Together," I echoed, even as the word tasted like poison on my tongue. "We're going to revolutionize everything."

"Exactly," she said. "And I want you by my side when we announce it tomorrow. At the press conference."

The entire world would be watching. Elara was positioning me as her partner, her ally in this massive scheme. And for everyone back home—Ace, Don, the others—it would seem like I had betrayed them. I knew what it would look like: Toby Valorian, standing beside the woman who represented everything we had fought against, the mastermind behind GovCore's greatest weapon, CoreCoin.

The next morning, I found myself on the stage. The bright lights of the press conference were blinding as I stood next to Elara, our hands intertwined. I played the part to perfection—smiling, nodding, as if I were in total agreement every word she said.

Across the globe, people were watching. Ace would be watching. And I knew exactly what he'd be thinking—*What the hell is Toby doing?*

Everyone who knew me, who knew about my involvement with Bitcoin, would think I had switched sides. I could already imagine their confusion, their disbelief, as they saw me on that stage, seemingly hand-in-hand with the enemy.

Elara finished her speech announcing the CoreCoin, and I could feel the eyes of the world's gaze upon us. She squeezed my hand, a gesture of triumph, thinking that she had won, that I was hers.

But little did she know, I was only beginning to play my cards.

As we walked off the stage, I caught a glimpse of the camera feeds imagining the confusion on the faces of the people who had once trusted me. The plan was in motion. Now, all I had to do was survive long enough to see it through.

## UNVEILING OF THE BIRTHRIGHT

As we returned from the press conference that evening, the night air was heavy with anticipation. Elara had been quiet on the drive back, but the tension between us was like a coiled spring, waiting to snap. When we finally arrived back at our luxurious home, she turned to face me. Her eyes gleamed with a mixture of ambition and something darker—something more dangerous.

"You did well today," she began. Her voice was soft, but there was an edge that suggested this wasn't a casual compliment. "You're getting closer to understanding where you belong, Toby."

I remained silent, watching her closely as she stepped toward me. "*Where* exactly do I belong?"

Her smile deepened. It was almost predatory. "At the top. With me."

She moved to a small table in the corner of the room, picked up a glass of wine and swirled it slowly, as if savoring the moment. "You're not just anyone, Toby. You're Paul Kestrel's son. Do you realize what that means?"

I looked directly into her eyes and kept my expression neutral despite the surge of confusion and curiosity in my gut. "I know the name, but what does it mean for me?"

Elara set her glass down and walked toward me. She spoke to me in a whisper. "It means everything. Power. Control. You could be the sole beneficiary of the largest segment of GovCore—the Collective. It's your birthright, Toby. You were never meant to be ordinary."

I swallowed hard, trying to process the meaning of her words. "But why now? Why didn't anyone tell me this before? And why would my father . . . why would he leave me out of it?"

She sighed, and her gaze softened, but was still tinged with that ever-present ambition. "My father, Paul Kestrel, had . . . a thing for women who weren't part of the GovCore elite. He enjoyed mingling with the lower classes, especially with women like your mother—smart, strong, but not bound by our system."

A surge of anger rose in my chest as I thought of my mother. "So, he used her?"

Elara hesitated, choosing her words carefully. "It wasn't just that. He couldn't bring you into the fold, Toby. You were . . . inconvenient. You were a secret he couldn't afford to have revealed. He was too high-profile. If anyone had found out, it would've jeopardized everything. So, he decided to have you eliminated."

Her words were a punch to the gut, but I didn't flinch. "Eliminated? You mean that's why you came to my home all those years ago?"

She nodded, and the cold, hard truth settled between us like a jagged blade. "Yes. But once you survived that first incident, once you avoided what they had planned . . . well, we realized you weren't a mistake. You were one of us. Part of the bloodline."

I could feel the ground shift beneath me. The reality of my past and the truth of my heritage pressed down. "So, what then? They just decided to let me live my life? No more attempts on my life?"

Elara took a step closer, and her voice became almost intimate. "Not exactly. We watched you, Toby. They let obstacles fall in your path, ensuring that you would never rise too high, that you would stay within the Collective. They allowed you to struggle to keep you from gaining real power outside of their control. But now . . . now, I want to bring you into the fold."

"And what about my mother?" I asked. "What happened between her and Paul?"

Elara paused and shook her head slightly. "I don't know all the details. My father never spoke much about it. But I do know that he was fond of her—more than any of the others. Something about her drew him in. Maybe it was her independence, her strength . . . but in the end, he couldn't let her disrupt the future he had planned. So he kept his distance."

I looked down, trying to absorb everything I had learned. My mother had been nothing more than a fleeting obsession to a man who had built an empire—and I was the inconvenient byproduct of that obsession.

Elara reached out, and lightly touching my arm. "But you're here now, Toby. You've made it through everything they've thrown at you. And now, with me, you can take what's rightfully yours. Together, we can reshape GovCore, rule it, and make it into something even greater."

I felt the pull of her words, the seductive lure of power and control. But beneath it all, there was ugliness, a betrayal that I couldn't ignore. I stared at her, seeing ambition flickering in her eyes, the same ambition that had driven her to this moment.

"Is that what you want, Toby? To rule?" she asked with a provocative tone.

I didn't answer right away. My mind was racing. The revelation of my father's past and my connection to GovCore swirled in a chaotic storm. There was a choice before me now—one that would change everything.

I forced a smile to mask the turmoil inside. "I'm ready, ready for it all!

Elara's smile widened. "It's yours for the taking. We have a future to build."

As I looked into her eyes, I knew one thing for certain: Whatever future she had in mind, it wasn't one I could trust. Not yet.

# CHAPTER 16:
# THE UNDERCURRENTS OF POWER

The next morning, after a restless night turning over Elara's twisted revelations, I woke to a different reality. I was still processing everything—my father's role, Elara's manipulations, and now, the weight of GovCore hanging around my neck like a yoke. But I couldn't dwell. I had to move.

Moving through the halls of GovCore's facility was like stepping into the belly of a beast. My security detail followed my every step, though I could feel their distance. I wasn't sure if they were protecting me or keeping an eye on me for Elara. The luxurious corridors were sterile, cold even. Every room reeked of wealth and technology, but something about it was suffocating.

Elara had given me access to the entire facility. Her hope, I assumed, was that I'd come to accept the power she had offered me. But as I wandered through the offices and labs, meeting the key players of GovCore, I could feel an undercurrent, a discontent brewing beneath the surface. It didn't take long before the whispers began to reach my ears.

The old guard—the people who had been with GovCore since its inception—were still there. Grizzled and cynical, they kept their distance from Elara, maintaining their positions but not offering her the loyalty she craved. Thesemen had known Paul Kestrel, who had built GovCore's empire, and there was something about seeing me—a male Kestrel—that seemed to stir something deep within them.

"You must be Toby Valorian," said one man as I entered a research wing. His name was Mason Eddard. He was an older man with silver hair and sharp eyes, a man who had clearly been around since my father's days. He shook my hand firmly, and his gaze lingered a little longer than was comfortable.

"I've heard a lot about you," he added, but he did not elaborate. Something about the way he said it made me feel that I was being tested.

Keeping a neutral tone, I said, "I'm still getting up to speed on everything. A lot has changed since my father's time, I imagine."

Mason gave a slow, calculating nod. "Yes, well . . . a lot has. And not all for the better."

I raised an eyebrow. "How do you mean?"

His eyes flickered, and his voice lowered a bit. "There are those of us who remember Paul. The way he led. It wasn't perfect, but it was different. GovCore was never meant to be . . . what it's becoming."

He didn't say it outright, but I could feel the emotion in his words. It was more than Elara's leadership he was referring to—things had changed since she took control. The Cerebrax Seal—this monstrous idea that everyone, eventually, would be fused to GovCore's artificial intelligence, linked to the mind of Neutrinon—it was a violation, even to some of the most loyal GovCore agents. It was one thing to rule with power and wealth, but the Seal was something else entirely. It was a form of control no one could escape from.

"I assume you're on board with the CoreCoin launch," Mason said, testing me. "Though I hear there's . . . hesitation in certain circles."

I gave him a tight smile, letting my words hang ambiguously. "It's a big step. Not everyone's ready."

Mason's expression shifted slightly. His eyes narrowed. "You know, many of us believe GovCore needs strong leadership. Not everyone thinks a woman like Elara is . . . fit for the job. Especially not with the direction she's taking things."

He made a direct shot with no attempt to hide his opinion. I glanced around, making sure none of Elara's loyalists were nearby. "And what do you think, Mason?"

He shrugged, but his voice was hardened. "I think someone with the bloodline of Paul Kestrel, someone with your background, should be calling the shots. Not someone trying to twist the legacy of GovCore into . . . into this."

I kept my face unreadable, but inside I could feel the currents shifting. It wasn't just Mason. The more people I met that day—the OGs, the scientists who had once worked under my father, the long-time operatives—they all said the same thing, though not in as many words. Elara's grip was slipping, and they were looking for an alternative.

They were looking to me.

By the time I returned to the lavish home Elara and I shared, I felt the reality of the day's conversations catching up to me. It was clear now that GovCore wasn't the monolith it pretended to be. There were cracks in the foundation, and those cracks were deepening. The old guard didn't want to be part of Elara's future, especially with the Cerebrax Seal looming over their heads. They feared it. And more than that, they resented her for it.

In their eyes, I was the heir apparent—not because I deserved it, but because I was male, and I had the Kestrel blood in my veins.

Elara might have been dangerous, but so was ambition. And now, I was surrounded by it.

I needed to play this carefully. One wrong move, and I'd end up at odds with both sides. But if I could manage to pit Elara's loyalists against the old guard, if I could create enough division within GovCore, I could set the stage for something bigger. Something that would tear this entire system apart from the inside.

As I sat in my room, Ace texted me, asking how things were going. I smiled at his optimism, though it was short-lived. I had told him I was going to handle this, but the deeper I got into GovCore, the more I realized how dangerous the game was.

I replied simply, "Everything's moving in the right direction. Trust me."

But inside, I wasn't sure if I could trust myself to get out of this alive.

The battle lines were being drawn, and I was standing in the middle of it all.

Chapter 13: The Genesis of Rebellion

After my long day of quiet conversations with the old guard, I knew the next step was crucial. I couldn't rush into this without fully understanding how to frame it. The idea of forming a group within GovCore—a faction within a faction—was risky, but I knew it was the only way to dismantle the growing power imbalance. *From the inside.* The old guard had been dropping hints all day, subtle nods of discontent, whispers of the past when GovCore was built on unity and strength, not control. They didn't trust Elara, and in a way, neither did I. Not completely.

But I needed her to trust me. That's where my edge lay—in her belief that I was still discovering my place, that I hadn't fully grasped the magnitude of my lineage or the power I held. She underestimated me. Her arrogance was a chink in the armor I planned to exploit. She thought she held all the cards, that I was merely another piece on the chess board that she could manipulate. But I had my own game to play.

Sitting in my quarters that evening, I mapped out my plan to create a coalition buried within the machine—GovCore+. Not a new system, but a revival. The old guard. Hidden voices, forgotten protocols, and encrypted loyalties that still pulsed beneath the surface. We wouldn't dismantle GovCore from the outside. We'd rewrite its future from within. GovCore Plus had to sound like a tribute to the foundations her father built, the legacy she desperately wanted to preserve. I'd need to present it as a solution, something that would strengthen her position and bring the old guard back in line, a way to honor their wisdom while keeping them under control. She would love that—the illusion of control.

I couldn't afford to let her see through me, not yet. If she sensed even a sliver of rebellion, she'd shut it down before it got started. I had to tread carefully, play to her ego, make her believe this was her idea, her victory. It was the only way to plant the seeds of dissent without arousing suspicion. And when the time came, I'd be ready to let those seeds bloom into something more—a quiet revolution from within.

The next morning, I would approach her, not as an adversary, but as her loyal partner, eager to unite GovCore under her leadership while I quietly plotted its division.

I found her having breakfast in our lounge which overlooked the manicured grounds of the GovCore estate. She sat by the window, eating and drinking orange juice; her expression was pensive as the sunlight glared over the horizon in the distance. The sight was almost serene, a stark contrast to the turbulence boiling within me.

She glanced up as I entered and gave an affectionate smile at my unannounced presence. "Toby, what's on your mind?" she asked. Her voice was smooth as velvet, but there was an edge of curiosity. She could sense I had come with a purpose.

I took a deep breath and tried to keep my expression relaxed. "I've been thinking about the future, Elara. About GovCore's future."

She smirked and leaned back in her chair. "Haven't we all?"

I stepped closer and lowered my voice as if sharing a secret. "I think we need to bridge the gap between the old guard and the new blood in GovCore. You've seen it yourself—some of the senior members feel isolated. Left behind by all the changes. They have experience, insights. They deserve a voice."

Her expression sharpened, and her eyes studied me, but she tilted her head, and her voice became relaxed, "Go on."

I took a breath, carefully choosing my words, and knowing that the next few sentences could make or break this entire plan. "The old guard . . . They've been feeling sidelined, disconnected. They remember how things used to be when GovCore was built on more than power—when it was about preserving stability and securing a future. But now, with all the technological advancements and the introduction of the Cerebrax Seal, the old guard . . . well, they're wary."

Elara's face didn't shift much, but I could sense the change in her energy. She was listening. "They think I understand them. I can be their voice, their bridge to you. But it's more than that. I think we need to bring them into the fold, into something that makes them feel valued again. They will fall in line; they'll actually fight for this vision you're building."

"And what do you propose, exactly?" Her words were measured, but there was a flicker of curiosity. I had her attention.

I leaned forward slightly, keeping my tone confident, like I was fully invested in her future. "We form a group—a council, of sorts. GovCore Plus. It's a way of bringing the old guard together with the newer generation, bridging the gap between tradition and progress. A space where they feel included, where their voices are heard. They need to feel like their values aren't being erased by the technology or the changes you're introducing."

Her eyebrows arched; her fingers tapped lightly on the table. "And you think this will keep them loyal?"

"I do. Loyalty comes when people feel like they're part of something bigger. If we position this right, they'll see GovCore Plus as their chance to guide the future, to adapt the old values to your new vision." I paused, letting my words sink in. "And with me leading it, you'll have someone who can speak their language, who can keep them grounded while moving them forward."

Elara leaned back in her chair. A small smile tugged at the corner of her lips. "You've thought this through, haven't you?"

"I have," I said, maintaining eye contact. "You know how much power I can wield with them. They already look to me because of my father's name. But now they see someone who's risen through adversity, who's embraced the new world. They respect that. If I can give them a way to be part of this future, they'll follow us."

For a moment, silence hung between us. I could see the gears turning in her mind, weighing the risks, the potential outcomes. This was the crux of it—getting her to believe that this move would benefit her as much as it would benefit me.

Finally, she set her glass down and gave a small, conceding nod. "Alright, Toby. I'll allow it. But I'll be watching closely. Don't give them any ideas about resisting progress. We need unity if we're going to succeed."

I hid my relief behind a grateful smile and nodded earnestly. "You won't regret it, Elara."

She smiled back. There was something predatory about her gaze. "I hope not. Don't forget, you're one of us now, Toby. We both want what's best for GovCore."

I nodded, playing my part perfectly. "Of course. We're in this together."

But as I left the lounge, my mind was already racing ahead. She had no idea that this group was going to become the seed of something much bigger. It wasn't about uniting the old and the new. It was about creating a space where those who were disillusioned with Elara's rule could come together, where we could discuss alternatives—where we could plan.

The next day, I set to work, moving through the halls of GovCore like a man on a mission. I approached the old guard, the men and women who had hinted at their dissatisfaction. Mason Eddard was one of the first I spoke to.

"We're putting together a group," I told him, keeping my tone casual. "GovCore Plus. A place where we can discuss ideas, bring the old guard together with the new. I think it's time we start sharing our perspectives."

Mason raised an eyebrow, and smirked. "And you think Elara's going to let you do this?"

I shrugged. "She's given me her blessing. She sees it as a way to keep things steady. To smooth over the cracks."

Mason's eyes gleamed with understanding. "She doesn't know what she's allowing, does she?"

I smiled, leaving the question unanswered. "So, are you in?"

As he clapped me on the shoulder, his smirk widened. "I'm in, kid. Let's see where this goes."

One by one, I gathered the others. A whispered word in the hallway, a quiet conversation in the shadows of the courtyard. Each time, the message was the same—a chance to be heard, to regain a sense of purpose. And each time, I could see the flicker of hope in their eyes, a hope that maybe, just maybe, GovCore could change.

What Elara didn't realize was that the more we met, the more those subtle divisions between us would deepen. Those who were loyal to her, who believed in the Cerebrax Seal and the CoreCoin, would begin to feel the distance growing between them and those who believed in something different—something closer to the original vision of GovCore, a vision that didn't involve total control.

By the time the first official meeting of GovCore Plus took place, I could see the lines being drawn. Elara believed she had me under control, that I was playing into her hands. But she didn't know that I had a plan of my own, and it was about to change everything.

I had to play my part a little longer. And when the time came, I'd be ready to turn the tables.

# CHAPTER 17:
# RESTORING THE ECHOES

In the weeks following Elara's reluctant blessing, my group began to take shape. The older members, once sidelined and dismissed as relics of a bygone era, started to gather in private rooms, away from the ever-present gaze of Neutrinon. These weren't formal meetings with agendas or presentations; they were gatherings where stories flowed like currents, pulling us into the past that GovCore had buried beneath its gleaming façade.

At first, the conversations were cautious, like men testing the strength of thin ice. They spoke of the early days—back when GovCore wasn't a fortress built on surveillance and control. It was a gathering of minds determined to stabilize a crumbling nation. Some of them were among the original members, men like Mason Eddard, a former Navy strategist, and Dr. Caldwell, a neuroscientist who had once dedicated his life to understanding the mysteries of the human brain. They talked about the purpose they had seen in those days, a purpose that had vanished beneath the layers of bureaucratic power.

One evening, Mason Eddard leaned forward and spoke in a voice that was low but steady. "You know, kid, back then, it wasn't about controlling people. It was about giving them a chance. America was on the brink. Banks were failing, unemployment was through the roof. We were brought together because we had the means and the will to rebuild." He paused and looked around the room as if expecting ghosts of the past to nod in agreement.

Dr. Caldwell chimed in. His expression was thoughtful. His fingers slowly traced an old scar on his hand. "There was a time when we believed we were doing something good. The idea was to protect the nation from collapsing, to build a structure that could withstand anything. But that structure became a trap. For everyone."

As they spoke, a vision of GovCore emerged that I had not seen before—a vision driven not by the hunger for power, but by a deep-seated desire to shield a nation from disaster. It was a different side to the beast, a side that had become twisted and corrupted as the years rolled on. I could almost picture those original meetings in dimly lit war rooms, where battle-hardened veterans and wide-eyed academics sketched out a blueprint for a new America. And at the helm was my father, Paul Kestrel—ambitious, visionary, flawed.

Mason's sharp gaze cut through the haze of memory. "The thing is, son, we never meant for this to become what it is now. We weren't supposed to have our hands in every pot, to make every choice for everyone. But when the power started to concentrate . . . well, it's easy to lose sight of the line between guiding and controlling."

I felt the truth of their words settle in my chest—not like a burden, but like a key turning in a lock. I saw how their original intentions had been noble, how they had wanted to steer the country through the storm. But along the way, they had let power consume them, let it evolve into something monstrous and unrecognizable.

And now, we stood on the brink of a new chapter, with the CoreCoin and the Cerebrax Seal threatening to drag the world into a new kind of tyranny—a digital leash for every citizen, controlled by Neutrinon's cold calculations. It was an era that would erase the last remnants of the GovCore they had once dreamed of.

The older guard began to look to me as a bridge to that lost vision, a way to bring back the balance that had been lost. They didn't say it outright—no one did, not yet—but I could feel it in their glances, in the way they listened when I spoke.

One night, after another hushed gathering, Mason caught me by the shoulder as the others filed out. His grip was firm; the lines around his eyes were deep from years of battles fought and lessons learned. "I know you're not your father, Toby. But there's a lot of him in you. Just . . . don't make his mistakes. Don't let her blind you to what's right."

It wasn't a glass of wine he held, but a simple metal flask, a relic from his Navy days. He offered it to me as a silent gesture of understanding. I took a sip; the burn of the liquor reminded me that trust, once given, came with its own kind of fire.

These men didn't know everything about my plans, but they understood enough—they knew that Elara's vision for GovCore wasn't theirs. And as our group grew, so did our resolve. Each story they shared, each insight into the original intent of GovCore, became a piece of the puzzle I was putting together, a puzzle that would show me how to take down the monster from the inside.

As the weeks passed, I could see the gap widening between those who believed in Elara's vision and those who believed in something different. It was subtle, barely noticeable to the untrained eye, but it was there—like the shifting of tectonic plates, the promise of an earthquake to come. And I knew that when the time came, the ground beneath GovCore would shake. All I had to do was be ready.

# CHAPTER 18:
# THE GATHERING STORM

The holiday season was closing in; the air already carried that familiar chill. GovCore headquarters had subtly shifted in atmosphere with quiet preparations for the festivities underway. But for me, the growing tension masked any sense of cheer. My plan was moving into its next phase, and there was no better time to bring Ace into the fold.

Elara had promised me that Ace would have a place at GovCore, a future within the organization as one of their top scientists. She thought it was a gesture to keep me loyal—perhaps she even believed it was a genuine act of trust. But I knew better. If I was going to navigate this power play and maintain my facade, having Ace by my side was crucial. More than that, I needed to ensure his safety, and the only way I could do that was to keep him within reach.

I brought the idea to Elara that morning. The two of us sat across from one another in her sleek office. She was going over plans for the upcoming CoreCoin press tour, but I interrupted with a new proposal.

"I want Ace brought on as a scientist," I said, meeting her gaze directly. "It's time."

She paused and carefully studied me. "You've been patient with this."

"I had to be. But now I want him here. It's better for everyone, especially for the work you and I are doing. I'll fly back to Atlanta to give him the news myself. It'll be my last chance to see the world outside of GovCore, and it'll be good for him to hear it from me."

Elara's lips curled into a soft smile. "Of course. You know I always intended to bring him in, Toby. It's part of the vision. If you want to go to Atlanta, then go. Take the jet. Bring your brother back into the future we're building together."

I nodded and hid my relief behind a calm expression. "Thank you."

It was almost too easy. She believed me completely now, she thought I was hers, thought I was committed to her plan. And maybe, in her mind, bringing Ace on board would tie me even closer to her. But that was the thing about arrogance—it blinded people to the subtle shifts happening around them.

With everything in place, I made arrangements for the flight. It would be my final time to step outside of the suffocating grasp of GovCore, the last time I'd breathe free air, see old faces, and remind myself of the world outside. And it was the perfect cover for the next step in my strategy.

The next morning, the GovCore jet was waiting for me. I boarded quickly, and the staff treated me as they always had—with a quiet deference that was starting to feel disturbingly natural. The flight was smooth; the hum of the engines blended into the background as I mentally rehearsed how I'd explain everything to Ace. He didn't know the full extent of my plans yet, but he trusted me. I needed to get him on board. I needed him to see what we were up against.

As the plane began its descent into Atlanta, I glanced out the window and watched the city skyline come into view. The sprawling streets of the place that had once been my home stretched out below, still unchanged in some ways, yet entirely foreign after everything I'd experienced at GovCore.

The wheels touched down on the tarmac, and the familiar buzz of Atlanta's Hartsfield airport filled the air as I prepared to leave the jet. I straightened my jacket, took a breath, and stepped out, ready to face Ace and set into motion the next stage of this dangerous game.

I hadn't told Ace I was coming. He would've been cautious, perhaps a little paranoid, and rightfully so. I knew GovCore had its eyes on us at all times, even though they believed I was fully on their side now. But it was better this way—catching Ace off guard, seeing his real reaction when I walked through the door.

After landing in Atlanta, I hailed a taxi to keep a low profile. It wasn't like the private jets and chauffeured cars at GovCore. It felt like another life—taking this normal mode of transport to our old apartment—blending into the ordinary world once more. As the car pulled up to the apartment complex, I stared at the familiar building with a sense of nostalgia mixing with the tension I carried from GovCore. I was about to step back into my old life, even if it was just for a moment.

I made my way up the steps. The key in my pocket was from the days before GovCore had completely consumed my existence. I turned it in the lock and entered quietly, the door creaked as I stepped inside.

Ace was there, hunched over a workbench filled with wires, gadgets, and tools. He looked up, and his eyes bulged as he saw me standing in the doorway. His face split into a smile, but it quickly shifted into something else—worry, maybe even suspicion. He knew the world we lived in. He knew what GovCore was capable of.

"Hey," I said. My voice was steady, but I could see hesitation in his eyes.

"Toby?" he asked cautiously, getting to his feet. He smiled again, but it didn't seem genuine. His hands twitched toward a device on the table, something modern and unfamiliar, but I caught the movement.

"It's me," I reassured him, but Ace wasn't taking any chances. I couldn't blame him. After all, I had just returned from the belly of the beast. GovCore had resources beyond our imagination, and he had every reason to be cautious.

Without saying a word, Ace lifted one of his gadgets and began scanning me. The small device emitted a faint blue light as it passed over my body. He looked at the readout; then when he looked back at me, his eyes narrowed.

"Just making sure you don't have the Cerebrax Seal," he said quietly. His tone was apologetic but firm. "You know they can infuse it without you even knowing."

I nodded, allowing him to complete his scan. I understood his paranoia. It wasn't mistrust—it was survival. After a few tense seconds, the scanner beeped softly, and Ace let out a sigh of relief.

"You're clean," he said and dropped the device on the table. "But you've been gone a long time, man. I didn't know what to expect when you walked through that door."

"I get it," I replied as I entered the room. "I don't trust them either. That's why I'm here."

Ace raised an eyebrow, waiting for me to continue. I could see the questions in his eyes—why now? why was I back? and most importantly, what had I brought with me from GovCore?

"I need you to come back with me to GovCore," I said in a firm voice. The request hung heavy in the air between us, but Ace didn't flinch. He folded his arms across his chest, already shaking his head.

"No way," he said. "I'm not stepping foot back in that place. You don't know what they're capable of—"

"I do," I cut him off. My tone was sharper than I intended. "I've been in there. I've seen it firsthand, Ace. I know what they can do, and that's exactly why I need you with me."

He was quiet for a moment. He studied me as he tried to piece together what I wasn't saying. "Why?" he finally asked. "Why would I go back there?"

"Because we're going to take them down," I said simply. His eyes got bigger, but I pressed on. "Look. They trust me now. They think I'm on their side. Elara . . . she thinks we're in this together, that I'm committed to her vision for GovCore. But I'm not. I'm playing the long game."

Ace frowned, still processing. "What's the plan, Toby? You don't just take down something like GovCore from the inside. They've got Neutrinon watching everything, and Elara... she's no fool."

"That's why I need you," I said and stepped closer. "I need you to create a virus. Something that can disable Neutrinon. But here's the thing—it can't be just any virus. Neutrinon's too advanced. You can't infect it while it's online. There are too many safeguards, too many sensors. We need to shut the power off first."

Ace's eyes lit up with understanding. "A reboot," he muttered. "The virus needs to hit the system when it's rebooting, when it's vulnerable."

"Exactly," I confirmed. "When the system powers back up, the virus will be in there, corrupting everything from the inside. It'll be undetectable until it's too late."

Ace ran a hand through his hair as he paced the room. "You're talking about shutting down the power at GovCore. That's not going to be easy, Toby. They've got backups upon backups. You'd need to create some serious chaos to distract them long enough."

"I've already started that," I said keeping my voice steady. "I'm forming a group within GovCore, the old guard, the ones who are disillusioned with Elara. They don't want the Cerebrax Seal. They don't want CoreCoin. They're on my side, even if they don't fully realize it yet. When the time comes, we'll create enough chaos to keep everyone distracted while you implement the virus."

Ace stopped pacing and stared at me in disbelief. "You've really thought this through," he said slowly.

"I've had to," I replied. "This is the only way. And I need you with me, Ace. You're the only one who can make this virus work. I can create the chaos, but you're the key to taking down Neutrinon."

For a long moment, Ace was silent. His eyes were distant as he turned the plan over in his mind. Then he nodded, and a slow, determined smile spread across his face.

"I've already been working on something," he said, walking over to his desk. He opened a drawer and pulled out a small, sleek device, and held it up for me to see. "I was planning to use it on you if you came back with the Cerebrax Seal."

I laughed and shook my head. "Good thing I didn't."

Ace grinned, but the seriousness of the situation returned quickly. "This could work," he said. His voice was more confident now. "If we can get Neutrinon offline, even for a few minutes, I can upload the virus during the reboot. It'll infect the system without them knowing."

"Then it's settled," I said. The importance of the moment sunk in. "We'll go back to GovCore. You'll create the virus, and I'll set the stage. We'll take them down from the inside."

Ace nodded. His expression was resolute. "Let's do it."

The next morning, after a long night of strategizing with Ace, I decided it was time to reconnect with the group. It had been too long since I had seen them, and with the plan starting to take shape, I needed to reassure them, let them know that everything was still under control on my end.

Ace was on board now and working on a scaled up version of the virus, and I had a growing support base within GovCore. But none of this would work without the help of the people I trusted most—the ones outside of the system who could help us when the moment came to make our escape.

I turned to Ace over breakfast; the remnants of last night's conversation were still on my mind. "We need to meet up with the guys," I said, breaking the silence. "Don, Patrick, and the rest of them. Let them know what's happening on the inside and give them reassurance that the plan's in motion."

Ace nodded and sipped his coffee. "You think they'll still be with us? It's been a while."

I smirked. "They've been with us from the beginning. They've been waiting for this moment as much as we have. I'll let them in on the plan, but I'll need their help when we're ready to move."

By early evening, we had arranged to meet with the group at a familiar spot— a discreet, private dining room at a local restaurant where we could talk freely. Don, the former Navy pilot, and a few other members of our Bitcoin group arrived one by one, each of them greeting us with smiles and handshakes. The tension of the world outside, the looming threat of GovCore, seemed to melt away in the warm glow of the reunion.

We ordered dinner, and after some lighthearted banter and catching up, I began to ease into the real reason for the gathering.

"Alright," I said, leaning forward. My voice was low but filled with purpose. "I know you've all been wondering about what's going on since I went back to GovCore."

They grew quiet, and their expressions were serious as they waited for me to continue.

"I've made inroads," I explained. "Elara Kestrel thinks she has me right where she wants me—next to her, fully aligned with her vision for the future of GovCore. But that's just part of the plan. I've been getting inside, making connections, finding the old guard who aren't happy with the way things are going. The ones who want to go back to what GovCore was supposed to be."

Don nodded. "So, you're saying we've got allies on the inside?"

"Exactly," I said. "And that's why I need you all to stay ready. When the time comes, when the chaos starts, we're going to need an extraction plan. We need to be ready to pull the trigger when the system collapses from within."

Patrick chimed in, "And what exactly does that mean for us?" his sharp ear listened for key details.

I took a breath. "Once we take down GovCore, Bitcoin will be the new currency. The digital dollar. No more paper money. We transition everything to digital. Bitcoin, Ethereum, and the other major cryptos will be the foundation of a new financial system."

The room filled with quiet murmurs. I could see the excitement in their eyes, the way they looked at each other, as if they could already taste the future I was painting for them.

"Bitcoin as the new dollar," Don repeated, and a wide grin spread across his face. "That's going to change everything."

"Exactly," I continued. "We're talking about a world where everyone can have a piece of it. No more centralized banks. No more control by a few at the top. It's decentralized. It's digital. It's secure. And because the U.S. will be at the forefront of it, Bitcoin will continue to rise in value. It'll be an investment in itself."

One of the other members of the group, a tech-savvy guy named Mike, leaned in. "And what about GovCore? You think they're going to let that happen?"

I met his gaze with a firm voice. "No. They'll fight it every step of the way. But that's why we're doing this from the inside. When they least expect it, we'll pull the rug out from under them. And when the dust settles, Bitcoin will be what's left standing."

The room was silent for a moment as they processed what I was saying. This was no small plan. This was about toppling the most powerful entity in the country and replacing its currency with one that belonged to the people.

Patrick raised his glass; he had a smirk on his face. "To the future," he said. "To the new digital world."

Everyone raised a glass, and for a moment, we were a group of friends again—sharing laughs, dreaming about what we would do with all the Bitcoin we had amassed, talking about the freedom it would bring us when this was all over.

But beneath the laughter and the camaraderie, there was a shared understanding of what lay ahead. This wasn't about wealth alone. This was about survival. This was about building a new future on the ruins of the old one.

As the night wore on, I made sure to have a one-on-one conversation with each of them, reassuring them of my commitment and my plan. They all agreed to help when the time came, to be the eyes and ears on the outside, and to stay low until we were ready to make our move.

When the dinner finally ended, I felt more confident than ever that we had a shot at making this work. These were the people I trusted most, and together, we were going to change the world.

Before we left the restaurant, I pulled Don aside.

"Keep your planes ready," I said. "When the time comes, we're going to need a quick exit."

He gave me a sharp nod. "You got it. Just say the word."

With our plans solidifying, Ace and I left the restaurant and stepped out into the crisp night air. It was almost surreal, standing there under the stars, knowing that the future of everything was hanging on the edge of a cliff.

"We've got this," Ace said quietly as we made our way to the car.

I nodded, though my mind was racing with thoughts of what we had set in motion.

"Yeah," I replied. "We've got this."

But even as I said the words, I couldn't shake the feeling that we were headed into a storm far bigger than any of us anticipated.

# CHAPTER 19:
# THE WINTER FAÇADE

As the sun dipped beneath the horizon, the plane touched down on the private GovCore airstrip, leaving behind streaks of fiery orange and deep purple in the sky. Snow, fake or not, dusted the edges of the landing strip, giving it a postcard-perfect winter scene. As we taxied to the terminal, I glanced at Ace, and he gave me a subtle nod.

"Ready to put on the masks again?" I asked. I knew we needed to step into our assigned roles from the moment we disembarked.

He gave a faint grin. "Ready as ever."

We moved smoothly through the private terminal wearing carefully neutral expressions. As soon as we exited the plane, we were greeted by GovCore staff—professional, silent, and efficient. A black sedan with tinted windows waited for us. Its exterior blended seamlessly with the controlled, precise environment that was GovCore.

But as we rode deeper into the headquarters, the landscape shifted dramatically. Holiday lights glittered from every rooftop and tree, casting an almost otherworldly glow against the sharp lines of GovCore's brutalist architecture. Towering fir trees—too perfect to be real—lined the walkways, decked out in metallic ornaments and strings of lights. A wreath, easily eight feet wide, hung from the entrance to the main building. Everything was pristine, as if an entire winter wonderland had been imported and meticulously arranged for show.

"She really went all out," Ace murmured, peering through the window at the spectacle unfolding around us.

"It's like Santa threw up on this place," I said, half-amused, half-wary. The excess felt out of character for Elara, who was more practical than sentimental. But I understood. This wasn't about holiday cheer. This was a spectacle, a show of power and perfection in preparation for something far larger.

We pulled up to the main building, and the transformation continued inside. Glittering garlands looped along every passageway; poinsettias sat on marble countertops, and decorative snowflakes dangled from the ceilings. The scent of cinnamon and pine wafted through the air, no doubt by courtesy of carefully calibrated scent machines. Staff bustled everywhere, setting up cameras, stringing lights, and placing monitors in every strategic corner.

GovCore was no longer only the headquarters. It was a production studio now, a stage set for a global event.

"Guess Elara's busy," Ace muttered under his breath when he noticed the pace of activity. Workers rushed from meeting rooms to conference halls, carrying tablets and blueprints. Their expressions were tense with purpose. It was clear that the launch of CoreCoin had consumed the entire organization.

"She's not just busy," I said, scanning the faces we passed. "She's out of touch. They've got her buried in this launch."

Ace nodded. His cutting eyes studied the sea of unfamiliar faces. "Good. That works for us."

While being escorted through the halls of GovCore, we passed rooms filled with executives, tech teams, and camera crews huddling over logistics. Large screens mounted along the walls displayed countdown clocks for various checkpoints in the production timeline. Christmas Day was circled in bold red letters everywhere.

"They're really doing it," Ace said. "Broadcasting the CoreCoin launch on Christmas Day. They've turned this whole place into a stage."

"Because nothing says holiday spirit like a new world currency," I muttered, feeling the weight—not on my shoulders, but pressing in from all sides. It was heavy, almost breath-taking knowledge that something diabolical was in motion.

We moved pass the Nexus buzzing with GovCore personnel working feverishly with Neutrinon. The artificial super intelligence was driving the entire operation, analyzing every possible glitch and contingency. From the center of the control room, Neutrinon's holographic interface monitored everything like an omnipresent eye.

"They've got Halix working overtime too," I said when I caught sight of him through a glass partition. His movements were precise and sharp. He directed teams like a general on a battlefield.

"Yeah, looks like she's got everyone running on fumes," Ace added. "This launch is their everything."

At that moment, it hit me how pivotal the next few days would be. Elara, Halix, and the entire GovCore apparatus were betting everything on this launch. The CoreCoin was more than currency—it was their ticket to total control. With it, they could track, tax, and manipulate every transaction made by anyone connected to the system. And by integrating it with the Cerebrax Seal, they could ensure that no one would escape their grasp.

The staff left us just outside the conference wing where Elara was no doubt holed up in endless meetings. I glanced at Ace. "Once we get inside, just play it cool."

"Cool as ice," he said with a small grin. "But seriously, how do you think she pulled all this off?"

I shook my head, still processing the scale of it. "She's good. But she's also trying too hard."

Ace tilted his head. "Why Christmas Day, though?"

I exhaled slowly. "Timing is everything. It's symbolic. They want it to feel like a gift. Something new, shiny, and unavoidable. A distraction wrapped in goodwill."

"And everyone's too busy with their holiday plans to notice what's really going on," Ace added.

I nodded. "Exactly."

The plan was beginning to crystallize in my mind. Elara thought she had us exactly where she wanted us—smiling for the cameras, standing by her side, looking like loyal soldiers in her parade. But this was more than a launch. It was the beginning of the end. And she didn't even see it coming.

Chapter 15: Cracks in the Foundation

As we stepped into the room where Elara waited, the energy shifted immediately. She was more cheerful than usual; her smile was genuine, but there was an underlying current of distraction—like she was juggling too many thoughts at once, too many responsibilities. She rose gracefully from her seat and came toward me with open arms. There was a glint of satisfaction in her eyes.

"You're back," she said; her voice softer than I expected. "I was wondering when I'd see you again." She turned to Ace, and her expression was still warm, but it was laced with curiosity. "And Arthur. Good to see you again"

Ace stiffened slightly but managed a polite nod. "It's just Ace," he corrected gently.

Elara smiled, amused. "Ace, then. Welcome to GovCore."

Her demeanor made it clear that she saw this as a victory. I was here, with my brother in tow, as she'd hoped. I leaned into the role I needed to play, stepping forward, radiating the confidence and warmth she was expecting.

"You've been working too hard, Elara," I said; my tone was low but intentional. "You should take a break. Let me take you to dinner. Let's unwind a little."

She blinked, surprised but pleased by my offer. "I suppose I could use a break," she admitted; her shoulders dropped slightly, as if my suggestion gave her permission to relax.

I smiled, the picture of a concerned partner. "Perfect. Let's head over to the dining wing. Jm,ust us."

As we made our way through the illuminated halls of GovCore, the buzz of preparations for the CoreCoin launch was all around us. Staff members flitted back and forth, setting up lighting rigs and screens. Workers adjusted cameras and prepared press kits. Every corner of the massive facility hummed with anticipation and the air was thick with the aura of the impending global announcement.

When we arrived at the fine dining wing, as expected, everything was elegant, quiet, and tucked away from the frenzy of the headquarters. White-clothed tables, low lighting. And an impressive wine collection framed the room. I pulled out Elara's chair, and she gave me a grateful smile as she settled into her seat.

We ordered, and the waitstaff served us with gracious efficiency. Over the course of our meal, Elara began to unwind further, and the fatigue of her responsibilities melted away under the influence of fine wine.

Leaning forward, she rested her chin in her hand. "You know, this CoreCoin launch . . . It's everything. It's going to change the world, Toby."

I nodded, playing my part perfectly. "Walk me through it. I want to know everything. How it's all going to play out."

Her eyes gleamed with excitement as she outlined the plan. "The global broadcast will happen on Christmas Day. Every world leader, every major corporation, every government official—all of them will be watching. Neutrinon will integrate CoreCoin into the existing financial structures. It will start as a voluntary shift, but within months, Bitcoin and every other cryptocurrency will be converted. Those who don't comply . . ." She trailed off, letting the implication hang in the air.

I masked my expression, and asked, "How will they convert Bitcoin holders?"

"The Cerebrax Seal," she said with a sly smile. "Anyone wishing to convert their Bitcoin will need it. It ensures compliance. No one gets through without it."

I tapped my fingers against my glass and pretended to admire the ingenuity. "And during the broadcast, how will it all unfold?"

"We'll address the world from the courtyard," she explained. "The announcement will be framed as an invitation to embrace the future. And I want you there with me." She moved in closer. "I want to introduce you to the world as Paul Kestrel's son. The rightful heir."

The words hit me harder than I let on. Elara had just revealed her trump card. She wanted the world to know my lineage—to cement my place by her side. I nodded slowly, pretending to accept the gravity of her offer.

"Of course," I said with a smile. "I'll be right there with you."

We finished our meal, and as we made our way back to our home, Elara grew more affectionate. She was clearly tipsy from the wine. She leaned against me as we walked; her body was relaxed; her guard was down. By the time we reached our private residence, she was nearly asleep.

I carried her inside and gently placed her on the bed. As I pulled the blanket over her, she stirred briefly, whispering, "This is only the beginning, Toby . . . We're going to change everything."

I smiled down at her, masking the churn of thoughts beneath my calm exterior. "Sleep well, Elara."

With her safely asleep, I slipped out of the room and retreated to the other quarters, where Ace was waiting. We locked the door and quickly launched a whispered conversation.

"She told me everything," I said. "The broadcast, the CoreCoin rollout, the Seal. It's all happening during the announcement."

Ace's eyes narrowed. "What's the plan?"

"We have another GovCore Plus meeting before the broadcast," I explained. "That's when I'll need you to meet with the old guard. We need to shut off the power—only then can we inject the virus."

Ace nodded. "Some of the guys in the group are engineers and scientists. They know the systems better than anyone. They'll help us pull it off."

I grabbed his shoulder. "While everyone's gathered for the announcement, you'll meet with the chosen ones in secret. They'll give you the blueprint. We cut the power during the broadcast. That's when you release the virus."

Ace grinned. "And by the time they power back up . . . Neutrinon will be compromised."

I gave him a confirming nod. "Exactly. This whole thing will crumble from the inside. We just need to play our roles a little longer."

Ace's grin faded as the implications of our plan settled over us. "You sure we can pull this off?"

I met his gaze. "We don't have a choice."

The two of us sat in silence for a moment feeling the enormity of what lay before us. But there was no turning back now. We were all in.

"Let's make history," I whispered.

Ace gave a small, determined smile. "Let's do it."

As the night deepened, I knew that the next few days would be the most critical of our lives. The countdown to Christmas—and the CoreCoin launch—had begun . . . and with it, the final phase of our plan to destroy GovCore from within.

## COUNTDOWN TO REBELLION

The morning of the GovCore Plus meeting arrived, and the energy around headquarters felt more electric than usual, like static clinging to the skin before a storm. As I made my way through the chaos toward the meeting room, I kept my posture calm and steady as I rehearsed the plan in my head. We were on borrowed time, and every move from here on out had to be flawless.

Just as I was about to round the final corner to the meeting room, Halix emerged from the shadows like a predator stalking prey. His presence was cold and deliberate; his infrared eyes scanned my face. He stepped into my path, blocking me with an almost mechanical stillness. His expression was unreadable—a blend of caution and something darker. It was as if the lines between human and machine had blurred beyond repair.

"I know what you're up to," he whispered. "You've always been clever, Toby, but Elara . . . she's three moves ahead."

I kept my expression neutral. I offered no reaction. "Is that right?"

He leaned closer. "She's planning to offer the Cerebrax Seal to all the GovCore staff. A gift, she calls it. But it's a test. Anyone who refuses . . ." He went quiet, but the silence hung between us like a guillotine. "If they reject it,

I've already been given orders. My team will neutralize them. No negotiations, no second chances. Immediate elimination."

My pulse thrummed in my ears, but I didn't flinch. His gaze narrowed. He studied me like a predator waiting for the slightest twitch of weakness.

"I won't act until the command is given," Halix continued. "That's set for Christmas Day. But don't kid yourself, Toby. The window is closing. Fast."

I gave him nothing—no hint of understanding, no confirmation of my own plans. "Thanks for the heads-up," I said casually as if we were discussing the weather.

He tilted his head, and a thin smile twitched at the corner of his mouth. "Be careful who you trust."

With that, he stepped aside, and his presence dissolved back into the shadows. I resumed my pace toward the meeting room. My thoughts raced through the implications of what Halix had just revealed. I knew now that we were on a razor's edge. There would be no second chance.

When I arrived at the meeting room, the sight before me hit like a tidal wave. The room was packed—far more than I had anticipated. Young and old, men and women, all gathered, with taut faces. The quiet hum of conversation buzzed as they whispered rumors about the Cerebrax Seal.

Before I could say anything, one of the older members stood and addressed me directly. "We've heard things, Toby. Whispers that the seal won't be optional. That they'll force it on all of us."

I took a breath and surveyed the room. Their eyes were wide with a mixture of fear and defiance. They had gambled by aligning themselves with me, and now they needed answers.

"The rumors are true," I said, my voice cutting through the murmurs. "The Cerebrax Seal will be offered to all of you during the CoreCoin launch on

Christmas Day. And those who refuse . . ." I paused, letting the weight of the truth settle ". . . will be eliminated."

The room collapsed into stunned silence. A few gasps. Murmurs of disbelief. Then there was silence as the gravity of the situation took hold.

"We're out of time," I continued. "We have only days to act."

All eyes locked on me, waiting for the next step. I could see it in their faces—the desperation, the simmering rage, the flicker of hope. This was the moment that would decide everything.

"I have a plan," I said as I stepped forward. "And I need all of you to make it happen."

With that, I motioned to Ace, who stood at the side. He gave me a brief wave and began unpacking the tools and gadgets he had brought. His movements were methodical. He placed a futuristic, black device on the table and pressed a button. A low resonance filled the room as the device scanned for surveillance equipment that might have been planted.

"Clear," Ace announced, nodding to me.

I turned back to the group. "The plan is simple, but it's going to require precision. On the day of the broadcast, while Elara and the rest of GovCore are focused on the CoreCoin launch, we cut the power. Everything—lights, surveillance, Neutrinon—goes offline."

One of the older engineers stood up. He was intrigued. "How do you plan to cut the power? The systems are fortified. They've got redundancies."

Ace grinned and pointed to a rolled-up blueprint. "I've studied the schematics," he said. "GovCore's power grid may be complex, but it's not infallible. There's a choke point in the backup system—an access panel beneath the eastern wing. If we take that out, the entire grid collapses."

"And that's where the virus comes in," I added. "Ace already has it ready. Once the power is down, we upload the virus directly into Neutrinon. When the system reboots, the virus will integrate itself into the core code."

Another member of the group, a scientist with salt and pepper hair, raised a hand. "What happens if the power comes back on before we finish?"

"We won't let that happen," I insisted. "We'll have engineers stationed at every critical point, ready to delay any manual overrides."

"We shut it down from the inside," he began, his voice calm but sharp. "Power grid first — we'll need a blackout window to inject the virus into Neutrinon. Without it, the AI's firewall will neutralize the code before it takes root."

He tapped the map. "Team One goes underground to the core relay station. That's where we trigger the outage. Team Two — that's us — we infiltrate the Data Spine while the systems are blind. We'll have six minutes, maybe less."

Heads nodded around the room.

"We'll place lookouts near all key junctions — loading bays, north ingress, and the glass bridge. Once the virus is live, Neutrinon will start shutting itself down from the inside, piece by piece. We'll mask the breach as a power malfunction... by the time they realize it's deliberate, it'll be too late."

Ace looked up. "Each of you will have a role to play. This isn't just a rebellion. This is a coordinated takedown."

The room buzzed with energy as the tension shifted from fear to determination. The old guard, the engineers, the scientists—they were all on board. They didn't need convincing. They wanted their world back as badly as I did.

I stood at the head of the table, looking out at the group. "This is it," I said. "We shut down the power. We release the virus. And we dismantle GovCore from the inside." I took a moment to breathe it all in. We had done it — gathered the minds, the resolve, and the plan. The conversation that night was long... and necessary. We went over every detail roles assigned, timing dissected, fallback contingencies mapped to the second.

A mutter of agreement rippled through the room. The plan was in motion.

"We meet again on Christmas morning," I said. "Everyone be ready. This is our only shot."

The room slowly emptied as people returned to their quarters, each feeling the enormity of the task ahead. As Ace and I stood alone in the room, he looked at me with a somber expression.

"You really think we can pull this off?" he asked.

I clasped his shoulder and met his gaze. "We have to. There's no going back now."

With that, we gathered our things and left the room. The countdown had begun. And as I walked through the corridors of GovCore, I knew one thing for certain—Christmas was a day the world would never forget.

# CHAPTER 20:
# CHRISTMAS IN THE EYE OF THE STORM

The morning was deceptively quiet. The corridors of GovCore were filled with the scent of cinnamon and pine, as if someone had tried to bottle the essence of the holidays and inject it into the walls. Wreaths adorned doorways; garlands looped around polished railings, and the subtle harmony of carols played over the sound system. Snow had been brought in from somewhere and created a winter wonderland just outside the dining hall. It was a picture-perfect holiday morning, designed to soothe, to disarm, to lull everyone into a false sense of security.

Inside the grand dining hall, staff and executives gathered. Everyone was dressed in their finest. Elegant gowns, tailored suits, and perfectly knotted ties. Elara Kestrel glided through the crowd like royalty; her presence commanded attention without effort. She was stunning, dressed in a shimmering red gown. Her hair cascaded in waves, and a smile on her face that was both inviting and dangerous. Her appearance was effortless, but something beneath her polished exterior reeked of what was to come.

Elara had mastered the art of presentation. Everything about her this morning was designed to appear warm, generous, and gracious. She portrayed the perfect host on the most joyous of days. But I knew better. So did Ace.

I glanced at my brother from across the table. He was stiff and wore a black suit with a bright, festive tie. The two of us exchanged a quick look—an understanding, a silent exchange that needed no words. He knew what today meant. This was more than Christmas morning; it was the endgame.

Elara floated toward us, balancing a glass of mimosa between her fingers, and I could see it in her eyes—the confidence, the certainty that everything was unfolding exactly as she had planned. "Toby," she whispered, "Merry Christmas." She leaned in and kissed my cheek; the scent of her perfume lingered like a soft warning. "Arthur," she smiled at my brother and said, "Welcome to the family."

Ace forced a polite smile. "Merry Christmas," he muttered, but I could see the tension in his shoulders. He was playing along, but beneath his calm exterior, I knew his mind was racing.

"Today's a big day," Elara beamed. "We've worked so hard for this, and soon, the entire world will witness the birth of a new era."

"CoreCoin," I announced and lifted my champagne-laced juice flute as if I were genuinely celebrating. "The future."

She smiled, satisfied with my words, and took a sip from her glass. Around us, the staff continued to mingle. Laughter drifted through the air along with the scent of holiday spices. Plates of breakfast pastries and fine-cut fruits filled the tables. The finest cuts of meat were arranged with artistic precision, and every table was set with glittering crystal and gold-rimmed plates.

But beneath the glamour and elegance, there was a pulse —something dark and electric, like the moment before lightning strikes. Everyone here wore smiles, but those who had whispered their doubts to me before—the old guard, the skeptics, the ones who knew that Elara's reign was a lie—held their expressions a little too tight, laughed a little too hard. They knew, as I did, that the storm was about to break.

Elara came closer. Her breath was warm against my ear. "You'll be right beside me when it all happens, Toby. We'll show them what true power looks like."

I smiled, but inside, my mind churned. I knew what she was going to do. Halix had warned me—Christmas Day was the deadline. This holiday spectacle was a charade, a distraction. Once the broadcast began, everything would change. The Cerebrax Seal would be offered like a gift, and anyone who refused it would be marked for the terminal directive.

I sat back, so I could study her. "And what exactly happens after today?"

Her eyes sparkled with amusement. "After today, there's no turning back. The world will belong to us."

A soft laugh escaped her lips as if we were sharing a private joke. I forced myself to laugh along, but my chest felt tight. My hand brushed against Ace's under the table—a quick, subtle gesture. We were still on track.

Breakfast continued. The murmur of voices filled the air, but my senses were on high alert. I watched every movement, listened to every word exchanged nearby. The old guard members were here, scattered throughout the room, blending seamlessly with the crowd. They looked at me from across the tables, nodding slightly—silently acknowledging that they were ready.

Ace leaned over to me. His voice was barely audible over the clink of silverware and glass. "The blueprint's ready. We know where to hit."

I nodded without looking at him, keeping my expression neutral. "Stay calm," I whispered. "We can't tip them off."

Elara stood and raised her glass to make a toast. The room fell silent as she smiled at the assembled guests. "To all of you," she began, "thank you for your dedication, your loyalty, and your vision. Today, we begin a new chapter in history—one that will change the world forever."

Everyone raised their glasses; the sound of crystal chimed in unison and filled the air. I raised my glass as well, though my mind was far beyond this glittering illusion.

As the toast ended, Elara's hand lingered on my shoulder. "Walk with me," she whispered, and I followed her through the maze of tables and out into the hall.

The soft click of her heels echoed against the marble floor as we strolled down the corridor. "Today's going to be perfect," she whispered. "And when it's over, you'll be standing beside me, where you belong."

Her words were laced with a strange kind of affection—a possessive affection. I knew that she believed she possessed me, but she was wrong.

We paused in front of a large window overlooking the snowy courtyard. Below, technicians and crew members were setting up for the global broadcast. Cameras, lights, monitors—all were meticulously arranged. It looked more like the set of a high-stakes television production than the headquarters of a technocratic empire.

Elara stared out at the scene below and smiled. "It's beautiful, isn't it?" she murmured.

I glanced at her; my expression was unreadable. "It's something, all right."

As she turned to me, her eyes gleamed with triumph. "This is just the beginning, Toby. When the world sees what we've built, there will be no going back."

I nodded slowly, playing my part to perfection. But in the back of my mind, I was already counting the hours. The broadcast would begin in a few hours, and the window for action was closing fast.

As we stood by the window, watching the final preparations unfold, a strange sense of calm settled over me. The pieces were in place. The old guard was ready. Ace had the virus. And soon, the world would witness the fall of GovCore.

Elara slipped her hand into mine. Her touch was warm and commanding. "It's almost time," she whispered.

"Yeah," I replied, squeezing her hand just enough to let her believe I was still under her spell. "Almost time."

But as I stood there, holding her hand, I knew the truth. This was more than Christmas morning. This was the beginning of the end.

## FINAL PREPARATIONS AND SILENT REBELLION

I'd just left Elara behind outside the dining hall, her eyes laced with something between fear and fire.

I moved through the lower corridor level off-grid walkways beneath GoveCore's polished skyline—toward a decommissioned transit hub, long forgotten by updated maps. That's where the OGs waited.

The last GovCore Plus meeting. The last time we'd sit at the same table and speak as a united front before we shattered the world as they knew it.

It wasn't about resistance anymore. This was something else—something necessary. A rebellion!

As I entered the room the air felt heavy, like before a storm. As the members of GovCore Plus gathered in the discreet meeting room, the gravity of what we were about to do dawned on us. No one spoke louder than necessary. Every movement, every word was deliberate. We knew that from the moment the broadcast began, there would be no turning back.

One of the old guard members—a retired engineer named Caldwell—walked in with boxes of shirts in tow. "Got a little Christmas gift for everyone," he said with a wry smile and a gravelly voice. He had decades of experience. He began handing out the shirts; each one was the same shade of deep crimson, subtle enough to pass as festive attire but distinct enough to serve our real purpose.

I pulled one over my head and smoothed the fabric. It fit perfectly. The others did the same, and as I looked around the room, I saw the meaning behind

Caldwell's gift: solidarity. If things went wrong, we'd know exactly who to trust in the chaos.

Ace nudged me and tugged at his sleeve as a grin flickered across his face. "Nice touch," he whispered, tugging at his sleeve. "Almost festive."

"Almost," I said but the seriousness of the moment didn't escape either of us. This was no holiday party. This was war—war. Silent war in the shadows.

We gathered around the room's large conference table for one final walkthrough. The blueprint of GovCore's electrical systems lay in front of us, covered in notes and marked with red circles where key systems needed to be disabled. We went through it all again: When the broadcast started, Ace and a small group would slip away into the sub-levels and find the power control room. There, they'd execute the shutdown needed to inject the virus into Neutrinon's core.

Caldwell tapped the blueprint with a worn finger. "Remember, you only get one shot at this. When the lights go down, you have about two minutes to upload the virus before the backup systems kick in. If the AI senses the disruption . . ." He didn't finish the sentence, but the implications were clear. The AI-fused humanoids from Halix's elite team—would be on them in seconds.

Ace leaned over the table with his brow furrowed. "I'll be ready. I've tested the virus every way I know. We just need to get the power down long enough."

I nodded and placed a hand on his shoulder. "I trust you. We've got this."

One of the newer members of the group—a GovCore technician—stepped forward. "We've also secured access to the armory," he said, opening a duffle bag filled with guns. "We distributed firearms to the key members. I informed everyone. "We must be careful, if those robots sense hostility, it's over."

There was a collective silence in the room. Everyone knew what was at stake. If the AI picked up even the slightest sign of aggression, Halix and his team would descend on us without mercy.

I cleared my throat and looked around the table. "Listen," I said, "I won't be with you when all this goes down. I have to stay visible. I need to keep Elara distracted, make sure she doesn't notice anything's off. I'll be up on that stage for the broadcast, standing right beside her. You won't be able to reach me until it's over."

They nodded, understanding the risk. They knew that from the moment I stepped on that stage, I was on my own.

"Stick to the plan," I said. "And stay quiet. If we pull this off, GovCore is finished."

Caldwell gave me a firm handshake. "We've got your back, kid."

One by one, the group dispersed, moving like shadows through the corridors, blending seamlessly into the crowd of GovCore employees preparing for the broadcast. I caught Ace by the arm before he could leave.

"You ready for this?" I asked him.

He gave me a determined look, though I could see worry in his eyes. "Yeah," he said. "We're as ready as we'll ever be."

"Stick to the plan," I whispered. "And no matter what, don't get caught."

He gave me a hug and clapped me on the back. "See you on the other side."

I watched him disappear into the crowd. His crimson shirt blended into the festive chaos around us. A strange sense of pride washed over me—pride in my brother, in this ragtag group of rebels who had come together for one final stand.

Taking a deep breath, I straightened my shirt and took the final walk back into the bustling halls of GovCore.

Elara was waiting for me at the entrance to the main courtyard broadcast stage; her silver dress shimmered under the lights. She smiled when she saw me, and her eyes bright with excitement. "There you are," she said, linking her arm through mine. "Come on, we've got a broadcast to launch."

I smiled back, playing my role. "Right beside you."

As we walked toward the stage together, my heart pounded in my chest. This was it—the moment we had been planning. The storm was about to break, and there was no turning back.

Somewhere beneath us, Ace and the others were slipping into the sub-levels, preparing to cut the power. The virus was ready. The weapons were in place. All we needed was the right moment.

I stepped onto the grand stage with Elara. The cameras were capturing every moment. From where I stood, I could see the entire courtyard teeming with people—loyalists, staff, engineers, and the old guard scattered throughout the crowd. They were waiting for the big reveal. CoreCoin was about to launch, live to the world, tethered to a system that would enslave anyone who dared resist.

Elara's fingers were intertwined with mine, to keep me close as she whispered promises of power and control. She thought she had it all figured out—the son of Paul Kestrel by her side, the world at her feet. But I kept my own promises buried deep beneath the surface, waiting for the perfect moment.

The broadcast started smoothly; the world watched as Elara introduced CoreCoin to the masses. Ace's voice buzzed faintly in my earpiece, giving me updates. The power was about to be cut. The virus was almost in place. But we had a problem—Halix was onto us.

## CAUGHT IN THE CROSSFIRE

Ace and the GovCore Plus team were moving through the final access tunnels. The virus was set to upload as planned. But just as they finished, Halix intercepted them.

I knew it the moment I heard his voice in my earpiece—cold, mechanical, merciless. "You've made a grave mistake," Halix said. "You won't leave here alive."

Ace stalled and answered Halix with measured words, but it wasn't long before the fight erupted. Blows were exchanged, shots fired—the quiet subterranean of GovCore exploded into chaos. Meanwhile, on the far end of the complex, Caldwell's squad moved through the maintenance conduits, rerouting through legacy access panels as they closed in on the main relay. They reached the core control and triggered the shutdown. Then, it happened. The entire system shut down. Everything. For just a moment. The lights died. The cameras went dark. Halix's enforcers collapsed. Their connections were severed. In that fleeting blackout, Ace's crew made their move—linking the drive, uploading the virus. In silence, the rebellion began to speak.

For one brief moment, we had won. The virus was in. The system was vulnerable.

*But that moment didn't last.*

## THE WAR BREAKS OUT

The battle spread in an instant. GovCore Plus members clashed with the loyalists—fists, gunfire, and makeshift weapons echoed through the halls. The old guard, engineers, and renegade scientists fought with everything they had, desperate to prevent the Cerebrax Seal from becoming a global nightmare.

I watched the broadcast unfold from the stage, aware that our carefully crafted plan was unravelling. Elara was still unaware, wrapped in her delusions of grandeur as the cameras stayed locked on us. Her hand squeezed mine, and her eyes glowed with triumph

Just as we thought we'd gained the upper hand, the power roared back to life. Lights blazed. Systems rebooted. Halix staggered to his feet; his movements were jerky, but functional. The AI wasn't completely shut down—Neutrinon was still fighting back.

On the central monitors, Neutrinon's digital face glitched and distorted, warping into fractal patterns before sharpening into focus with renewed aggression.

The lights flickered—a sign that the virus was spreading. But it wasn't enough. Not yet. Neutrinon was still online, and Halix was still in control.

Sparks burst from data nodes embedded in the walls. Ceiling vents howled as cooling systems overloaded. Static crackled across the comm channels, and then Neutrinon's voice cut through—cold, emotionless, synthetic.

"You cannot corrupt perfection."

Panels opened along the walls as dormant security drones stirred to life, their optics glowing an ominous crimson. Halix's body twitched again. His pupils constricted, and an eerie buzz vibrated through the floor beneath our feet— his neural sync with Neutrinon reestablishing in real time. He was reconnected. Recalibrated. Reawakened.

The war raged on. GovCore Plus fought tooth and nail, knowing the virus was still working its way through the network, inching closer to Neutrinon's core. But Halix and his enforcers were relentless. Their cold efficiency had us on the brink of collapse.

Every hallway had become a choke point. Smoke coiled through shattered glass doors. Bullets and plasma pulses shredded the silence. Bodies hit the floor—some human, some fused with steel. Still, we pushed forward. We had to.

My heart pounded as I watched the chaos spill into the courtyard. Elara sensed something was wrong—her grip on my hand tightened—her smile finally faltered.

Above us, Neutrinon's face stretched across every screen, every reflective surface, as if the very infrastructure of GovCore were its nervous system. The virus was working, but Neutrinon was resisting. Its code had evolved. Patches were self-healing. Countermeasures were deploying in real time.

Ace's voice crackled in my earpiece. "We're almost there. But we need more time. Hold them off."

More time. That was the one thing we didn't have.

## A TURNING POINT

Ace's voice crackled in my ear. "Toby, it's close. Just a few more seconds."

I gave Elara a reassuring smile to mask the storm brewing inside me. This was it. Everything hinged on the next few moments. If the virus reached Neutrinon, Halix and his army would fall. If it didn't . . .

Well, there wouldn't be a second chance.

## THE COLLAPSE OF CONTROL

Then, just as Halix lunged toward Ace, the final piece of the virus hit Neutrinon's core.

Halix froze mid-step with an expression flickering between fury and confusion. The enforcers crumbled; their systems crashed beyond repair. The loyalists faltered; they were disoriented without their AI counterparts to guide them.

And just like that, the tide shifted.

GovCore Plus surged forward and overtook the loyalists in a sweeping victory. The fight wasn't over, but the balance had tipped irreversibly in our favor.

## THE AFTERMATH

I stood beside Elara as her world crashed down around us. She knew. She knew I had betrayed her.

Her smile was replaced by a cold, dangerous glare. "What have you done, Toby?"

I didn't answer. There was no need. She knew.

# CHAPTER 21:
# WAR FOR GOVCORE—THE FINAL DESCENT

The courtyard was filled with confusion and the tension of battle. Neutrinon's cold systems glinted as the virus continued its slow crawl through GovCore's infrastructure. The sounds of distant gunfire and shouting blared through the headquarters as GovCore Plus fought against the remaining GovCore enforcers. Every moment brought us closer to liberation or complete destruction.

Elara remained by my side, her demeanor unnervingly calm! I kept my eyes locked on hers, knowing that despite the seductive charm she had shown me earlier, a deeper menace lurked beneath the surface. Then everything cracked open. The stage flooded with GovCore loyalists; security and enforcers trapped me, rifles raised. Their black uniforms shimmered under the sterile lights.

I moved on instinct grabbed Elara, twisted her arm behind her back, and yanked her close. With my free hand, I drew the gun from the small of my back and pressed it into her side.

"Back off," I shouted. "Or she dies."

"The guards froze. Behind me, GovCore Plus fighters stormed the stage, gunfire erupting in every direction. Chaos exploded—fists, bullets, and bodies colliding in a brutal struggle.

As I backed into the crowd Elara didn't fight me. If anything, she seemed strangely amused. Her eyes met mine, and she smiled as if to say, "You always had it in you."

GovCore Plus was holding its own. The virus was dismantling Neutrinon, CoreCoin and triggering glitches across the facility.

"Call them off," I growled at her and twisted her arm tighter. "End this."

She gave a soft, breathless laugh. "Oh, Toby. You think you've won? You don't know GovCore, not really."

Her words sent a chill through me, but I kept my grip firm. "Try me."

She turned closer her lips touching my face and whispered: "I'm pregnant."

I froze.

The world blurred. Her words cut through the violence like a bullet. The night we had spent together, wrapped in fleeting intimacy, now came back with brutal clarity. It wasn't just her ploy to control me—it was real.

I stood there, stunned, trying to make sense of it. For a moment, everything fell away: The gunfire, the fight, the rebellion, were lost in the weight of this revelation.

"What?" I whispered. My grip loosened for a second.

"It's yours, Toby," she purred. "Our child. You can't do this now. You can't destroy what we've started."

In that moment, I faltered. I released her arm. Just enough.

And that's when everything shifted.

With fluid precision, Elara slipped out of my grasp. Her small team of loyalists moved in, fast and practiced, extracting her from the chaos like a well-oiled machine. She gave me one last look—half victory, half sadness—and disappeared into the shadows, leaving me there with nothing but confusion.

Then I heard the alarm.

A deafening klaxon rang out through the premises, and the lights dimmed as red emergency strobes flashed in rhythmic pulses. The ground trembled beneath my feet.

"What the hell is that?" Ace's voice crackled over the comms; panic and anger cut through the noise.

"They're triggering the failsafe," I shouted back. "They're going to destroy the entire facility."

Everything made sense—GovCore's final contingency plan. If the headquarters were ever infiltrated or compromised, a sequence would activate to wipe everything out: data, people, and infrastructure alike. Elara was leaving us to burn.

"We need to move. Now!" I barked, rallying the remaining members of GovCore Plus.

The virus had done its part, disrupting key systems, but we were out of time. The self-destruct mechanism had been set in motion, and there was no stopping it now.

Ace was already working on an escape route. "I've got a plane fueled and ready on the east platform," he said over the comms. "We need to get there before this whole place comes down."

"What about the others?" one of the older members asked, glancing back toward the battleground.

I hesitated.

Some of the old guard—the ones who had fought alongside me—wouldn't make it out. They knew this. They'd made their peace with it. They'd chosen to stay behind and hold the line, buying us the time we needed to escape.

"They knew the risks," I said; my voice was heavy with regret. "We honor their sacrifice by getting out of here alive."

The ground shook more violently. The ceiling groaned under the strain, and cracks snaked across the walls. Fires were breaking out in pockets across the facility.

"We have to go!" Ace shouted.

We moved as fast as we could, navigating through the crumbling hallways as GovCore around us. The sounds of battle faded into the distance and were replaced by the eerie groan of a dying building.

Enforcers were still out there hunting us, but they were slower, less coordinated. Every second counted.

We reached the east platform just as the final tremors shook the ground. Smoke billowed through the air, and the glow of distant fires cast an orange hue over everything.

The plane was waiting for us. The engines roared. Don, the Navy pilot, stood at the ramp, waving us forward.

"Get on!" he shouted. "We don't have much time!"

We boarded in a frenzy; adrenaline pumped through our veins. As the last of us climbed aboard, Don hit the throttle, and the plane roared to life, lifting off the ground just as the facility behind us began to collapse.

I stood by the window, watching as the heart of GovCore crumbled into dust. Fires raged. Explosions rocked the earth. And somewhere in that chaos, Elara was gone.

For now.

We climbed higher into the sky, leaving the wreckage behind; Ace sat beside me, breathing hard.

"What now?" he asked. He sounded exhausted.

I stared out at the horizon. Reality was setting in.

"We rebuild," I said quietly. "We make a world where they can't control us. A world where Bitcoin isn't just currency—it's freedom."

Ace nodded, and a faint smile formed on his lips.

But deep down, I knew this wasn't over. Not yet. Elara was still out there, carrying my child. And with the fall of GovCore, the real fight was only just beginning.

The Bitpocalypse wasn't an ending.

It was just the beginning.

www.ingramcontent.com/pod-product-compliance
Lightning Source LLC
Chambersburg PA
CBHW070018120726
47909CB00003B/982